ABOUT THE AUTHOR

Dr. Tapati Mukhopadhyay is Reader in Geography in Siddharth College of Arts, Science and Commerce, Bombay and Recognised Postgraduate Teacher and Ph.D. Guide in Geography, University of Bombay. As a serious researcher, she is involved in several research projects. She is Director of a research project on Pattern of Wholesaling in Bombay, sponsored by the Indian Council of Social Science Research and is Co-investigator of a project on Indigeneous Tradition of Indian Navigation, sponsored by "Council of Scientific and Industrial Research". She is also a Member, Executive Committee, Indian National Cartographic Association (INCA) and Chairperson of the INCA Commission to study the status of Cartographic Education at Schools. She has been a Member, Advisory Committee on Coastal Management Regulation, Govt. of Maharashtra. Recently, she has been selected a Member of the Senate, University of Bombay.

COMMERCIAL GEOGRAPHY OF A METROPOLITAN CITY

Spatial Structure of Retailing in Bombay

TAPATI MUKHOPADHYAY
Siddarth College of Arts, Science and Commerce
Bombay

Foreword by

Prof. B. Arunachalam

Former Head, Department of Geography
Bombay University
Bombay

CONCEPT PUBLISHING COMPANY, NEW DELHI 110059

ISBN 81-7022-564-7

First Published 1995

Published and Printed by
Ashok Kumar Mittal
Concept Publishing Company
A/15-16, Commercial Block, Mohan Garden,
New Delhi-110059 (India)

Dedicated

to

The Memory of My Father

The Late Tara Prasad Chakraberty

who would have been

the happiest person to see this book

FOREWORD

The present work is a condensed version of a thesis titled "Retail Commercial Structure of Greater Bombay: A Spatial Analysis" submitted for a doctoral degree of the University of Bombay. It is the outcome of a painstaking research analysis following a gruelling task of compilation of primary data in the field, that sets a standard for similar field-based enquiries at micro-level.

Retailing studies have been carried out in the Indian context and elsewhere by economists, sociologists and scholars of commerce. But such studies in a spatial perspective to evaluate the spatial processes and order that underlie retail function in the urban, metropolitan situations in India are by and large lacking. Mrs. Tapati Mukhopadhyay, in this work, has undertaken a meticulous analysis of the spatial structure of the retailing function in Greater Bombay, and its hierarchic evolution in relation to the commercial urban growth of the metropolis and its dynamic socio-economic milieu. This study is of a pioneering nature, in that it examines the formal retailing landscape evolution within the overall urban functional land use mosaic of an Indian city, that is no doubt the nearest to the western parallel. The study brings out the sharp contrasts between the retailing character and hierarchy in the Indian and western cities.

Mrs. Mukhopadhyay had to break new ground to discover a proper methodology to deal with the enquiry. Compiled secondary data relevant to retailing, at micro-level, with pertinent locational co-ordinates are not only meagre but often inconsistent and error-prone, demanding frequent field-checks. Nor are they available in the required format for a spatial investigation. It is indeed commendable that the author could carry out this gigantic task of primary data collection and mapping in the field with a remarkable tenacity and perseverance to capture in totality all the observable features of reality. The work would act as a basic reference for future researchers in the field of urban retailing activity.

The work deserves a special mention for the cartographic rendering of retailing related parameters for a visual depiction and spatial analysis. The maps are well supported by tables and graphs.

A constraint of this study is that it leaves out of consideration retailing in the informal sector. Painfully aware of this limitation, the study had to keep out this aspect, which in itself deserves a full scale investigation by a team of workers. In the obtaining scenario, it is not even possible to estimate the share and role of informal retailing in Greater Bombay, except to state, in broad terms, that it is sizeable, related to the commuting behaviour of the pedestrain traffic, even more than formal retailing.

Similar studies of the larger urban and metropolitan retailing landscapes in other Indian cities would aid in building a South Asia urban retailing model, and help in appreciating greatly their 'Asian' character.

B. Arunachalam
Former Head
Department of Geography
Bombay University, Bombay

ACKNOWLEDGMENTS

In carrying out this research project I have gratefully received generous help and encouragement from so many persons and institutions that it is not possible to individually name all of them.

However, I must mention in particular that I have been fortunate enough to have the privilege of carrying out his research project under the guidance of Prof. B. Arunachalam, whose deep insight and understanding of the city of Bombay, and logical and scholarly approach to the subject, have greatly benefited in achieving quality and high standard of innovative approach to the study of the fascinating theme of Bombay's commercial landscape. I take this opportunity to express my sinceremost gratitude to him.

Eminent geographers of the stature of Prof. C.D. Deshpande and Prof. V.L.S. Prakasa Rao have not only granted me the privilege of having insightful academic discussion with them on several occasions, but also encouraged me to carry out this research project. I would like to gratefully acknowledge my debt to them.

An empirical study of retailing function in Bombay's commercial and spatial context like this has to depend not only on the primary and published data but also on the access to unpublished official data and maps. I am deeply indebted to Mr. D.M. Sukthankar, the then Municipal Commissioner, Bombay Metropolitan Corporation (BMC) and later on, Revenue Secretary, Govt. of Maharashtra for granting me access to unpublished data from the B.M.C. and land record maps from the Land Record Office.

Many officials of the B.M.C. (particularly, Department of Shops and Establishments and Department of Planning and Design), the Land Record Office, the Bombay Metropolitan Regional Development Authority (BMRDA) etc. also extended co-operation and help. I am very thankful to all of them.

This research project could not have been smoothly completed without the grant of Teacher Fellowship under the Faculty Improvement Scheme by the University Grants Commission for one academic year

(1985-86) and study leave of the same duration from Siddharth College of Commerce and Economics, Peoples' Education Society.

I would also like to thank Mr. Prasad Gogte for rendering very useful cartographical assistance and Mr. T.S. Venkiteswaran for typing.

Finally, I must acknowledge the contribution made by my husband, J. K. Mukhopadhyay, an Economist, at various stages of this research project. My little son, Somshuvra, also has shown remarkable understanding and co-operation.

Tapati Mukhopadhyay

CONTENTS

LIST OF TABLES

LIST OF FIGURES

LIST OF MAPS

1

INTRODUCTION
The Study Theme

1.1 Preamble : Bombay the Commercial Metropolis

Bombay, the "Urbs Prima in India", is considered to be the commercial and financial capital of not only Western India, but also of the whole country. This commercial pre-eminence of Bombay has its origin in the colonial trade of British India. As the Gateway of India on the western seaboard, overlooking the Arabian Sea on the short Suez route, the port city was the main trading outlet for the commercial exports of Western India, especially cotton and mineral ores and the landing point for assembled and processed Eurpoean goods during the colonial era. But over time, the purely commercial function got, on the one hand, strengthened by export of a range of raw materials and mineral ores from a vast hinterland extending upto Srinagar, Delhi, Kanpur, Nagpur and interior Deccan, and on the other, diversified, with the introduction of the cotton textile industry in the second half of the 19th century, following the introduction of the trunk rail arteries connecting Bombay with the agriculturally productive interior. Further diversification of industries followed after the First World War. Since Independence, the added functions of business and finance, as a sequel to the diversion of the port functions of Karachi to it, besided administration, led to a rapid rise in the primacy of Bombay. The city grew in number as the urban functions reinforced each other and this only led to further accretion of more economic attributes. Growth in number necessitated an inevitable spatial sprawl, which, on account of the insular nature of the city, directed growth primarily northward into the adjoining island group of Salsette, while adding to its intensity of growth by a vertical expansion in the older areas nearer the core. The diversification of functions over time did not bring down the relevance and importance of the commercial function of the city. In fact, the commercial function further grew and started developing complexities of intra-urban spatial differentiation and diversification to meet the dictates of the changed and changing urban and metropolitan environment.

In terms of its share in the national economy, Bombay accounts for over 10 per cent of the factory employment and more than 20 per cent of the joint stock company. The importance of the city also grew due to the development of Bombay High where a significant quantity of indigenous oil is being produced.

Bombay has a sizeable production of chemical fertilizers, a wide range of synthetic fibres, rubber, plastics and pharmaceuticals. The textile industry has been until recently, one of the mainstays of the city's employment and accounts for a major share of the national textile output. It produces in some form almost the entire range of products in light and medium engineering industry. Bombay city is the country's largest commercial film production centre.

Bombay mobilizes almost 20 per cent of the all-India bank deposit, and accounts for nearly 13 per cent of the total bank credit. The Reserve Bank of India, the State Bank of India and many other banks as also several national-level financial institutions like the NABARD, the IDBI, the LIC etc. have their headquarters here and so is the country's biggest stock exchange. Bombay contributes over 30 per cent of the central revenue collected by way of income tax, union excise duties and customs duty. The Western and Central Railway zones have their headquarters at Bombay which also have a network of a suburban rail system catering to a large working population of the city.

Bombay handles almost a quarter of India's overseas trade. The premier airport in the country is located in Bombay which services a very large number of international and domestic flights. Bombay accounts for 15 per cent of telephone connections and 25 per cent of the telex connections in the country.

Bombay's population as per 1991 Census is 9,908,547, contributing 12.5 per cent of total population of Maharashtra; out of this population about half are migrants; significantly enough, over half of these migrants have come from various other States and not from merely the States contiguous to Maharashtra. It has a wide trade hinterland stretching from Kashmir in the north to Belgaum and Hyderabad in the south and south-east and Kanpur and Nagpur in the east.

While the pre-eminence of Bombay in the world of commerce is thus seen, as to what extent Bombay can still be called the commercial metropolis and to what extent the intra-urban commercial development contributes to the overall commercial landscape and to what extent the industrial function has affected the commercial status are matters that are worthwhile an indepth investigation. The present study thus is an attempt to examine in detail the spatial structure of one dimension of the commercial function namely, retailing, that serves the consumer needs of the city population.

2.1 The Nature of the Present Study

Commerce encompasses a wide range of tertiary activities including wholesaling, retailing, warehousing, business functions, finance, commercial transportation and the like activities that are intricately interrelated and mutually sustaining under the common umbrella of urban external economies. The gambit of the whole commercial function is too wide a field for investigation even for a single city. In the present investigation, the study is confined to one segment of commerce, namely, the geographical analysis of retailing trade in the city, and to examine the other aspects of commerce only to the extent of examining their relation to the retail trade function.

2.2 The Objectives of the Study

The main objectives of the present study are:

(i) To examine the nature of temporal and spatial growth of retailing trade in Greater Bombay, in mutual relation to the demographic and functional growth of the city in general and in relation to commercial activity in particular;

(ii) To examine and evaluate the nature and intensity of spatial organisation of the retail trading activity of Greater Bombay;

(iii) To analyse the spatial differentiation and specialisation of retailing function and to search for an explanation, as related to the wholesaling function, and other specific urban parameters such as commuting pattern and residential functions;

(iv) To identify the intra-urban retailing hierarchy, if any, and its spatial characteristics; and

(v) To examine the relationship obtaining between retailing function and socio-economic correlates of the consuming population and to draw valid empirical inferences of urban retailing characteristics that can be tested in similar Indian urban situation leading to worthwhile generalisation.

3.1 Limitations of the Study

The study primarily deals with the retail activity in the formal sector and leaves out of consideration the functions in the informal sector for reasons of a comparable data base.

However, it cannot be denied that in the Indian urban context, retailing in the informal sector in the form of door-to-door hawking and unauthorised pavement shopping in shopping districts account for a sizeable proportion of the sales out-turn, employment and service dependence. In the absence of reliable statistics, it is not even possible to make an estimate of the relative share of informal retailing to total retailing.

It can perhaps be stated at this juncture that growth of informal retail sector in the Indian metropolitan scenario, like that of Bombay, is closely related to the formal retail activity and quite often is seen to be supplementary spatially. Since informal retailing shares many of the spatial attributes of formal retailing, the confining of the present study of formal retailing is not likely to vitiate the observations and findings by and large.

A practical problem experienced during the conduct of the study also needs to be noted here. There is a lack of separation of retail and wholesale data in office records. Hence, in the preliminary stage of usage of secondary data analysis, the study has concerned itself with the commercial function as a whole (except for floor space date investigation). However, field observations reveals that pure wholesale activity is limited mainly to grain, sugar and iron and steel. In all other commodities even within the wholesale bazaar-zone, wholesaling is mixed with semi-wholesaling and retailing. This lacuna of lack of distinct retail data base has been mostly overcome through primary data, gathered through intensive field-work.

Retailing being a consumer service-oriented function, it has, in a practical sense, a wide spatial diffusion, reaching even the remote interior of residential area. Unlike western urban centres where all retailing is in shopping zones, in Indian towns and cities, retailing is widespread and shops are present almost everywhere. But, for the present study of retailing in Bombay the analysis has been very largely confined to clusters of retailing that have been identified through field-work, as these are the areas which account for the lion's share of retailing in any residential district.

4.1 Data Base

The investigation is based both on secondary and primary data sources. While the economic classification of the census of Greater Bombay was used for a preliminary demographic spatial analysis of the population engaged in commerce, data on retail employment, floor space data for various commercial functions and residential functions have been collected from the unpublished records of the Bombay Municipal Corporation. Other major secondary data sources include the Bombay Metropolitan Regional Development Authority, Bombay, Central Road Research Institute, Delhi and Operation Research Group, Baroda.

Initial field-checks indicated lacunae in the secondary data base. It was imperative that extensive field-work had to be undertaken to cover a wide range of areas of concern for the study.

Field investigation was in different phases and depths of intensity. The entire Greater Bombay complex was investigated through a broad field study for identification of retail cluster areas. More detailed field accompanied by field investigation was undertaken to evaluate the spatial patterns and structuring of the retailing clusters. Quite an indepth field investigation was carried out in six retail clusters chosen for study of internal differentiation and characterization. Finally, the greatest depth of field anaylysis was resorted to in case of two retail clusters, namely, Fort and Mandvi-Bhuleshwar-Kalbadevi at a plot-to-plot level to study the horizontal and vertical characteristics and retailing morphology of the two areas.

The data collected from secondary sources were tabulated on the basis of Census section and wards and the primary data have been tabulated on the basis of retail cluster zones. The Census section map is used as a base map for final mapping. For detailed case studies, the municipal ward and section maps were used as base maps.

4.2 Techniques of Analysis

A discrete use of tabulation, descriptive statistical measure graphs, and curves as well as mapping has been made to analyse the data. More specifically, for analysis of occupational structure and functional character, location quotient method is used to find out the relative strength of each function. Mapping through choropleth method has been used. The nature of concentration of retail establishments is quantified by constructing cumulative curves, Lorenz Curve and dendograph. For determining the retail hierarchy, weightage score method, centrality index method, Guttman scaling method, retail strength method are applied. A composite scoring method is applied for determining the socio-economic hierarchy of Bombay.

Correlation and regression analyses are carried out for determining the nature of relationship between the retail hierarchy and city structure, population density and socio-economic status.

2

STUDIES IN GEOGRAPHY OF RETAILING: A Review

2.1.1 Retailing Function in Population-Space

Retailing is the final link in a chain of supply that begins with the growers and producers of goods and ends with the individual members of the buying public who are consumers of finished goods.

2.1.2 Studies of Retailing

The study of retailing in terms of its geographical perspective is comparatively recent. Although many economic studies of retailing have been made, the spatial organisation of an efficient retail marketing system to satiate the need of the consumers in terms of the commodity, the time and place where they require and the prices they are willing to pay, demand a geographical explanation that has not received as much attention as economic studies. In fact, the form, function, inter-relations that retailing organisation offers in space opens a fascinating field for worth-while and fruitful studies.

For many decades, the study of the retailing sector of the economy has been an area of keen interest of geographers. Various studies on retailing have also been undertaken by geographers as a part of analysis of urban functions and urban morphology. Pioneering works on retailing have been done by Proudfoot (1937)[1] and Hoyt (1939)[2] in America and Christaller (1933)[3], Smailes (1944)[4] and many others in Europe. Many of these studies were concerned with marketing among rural communities like that of Skinner (1964)[5] in China. All these studies have shown retailing as a major component of the urban environment.

Most of the early studies on retailing were essentially historical, descriptive and qualitative in nature. Geographic literature has plentiful studies of periodic markets in the Asian, African and European contexts. But organised formal retail market development in urban setting was much later to develop. Also, the later studies reveal a marked shift away from descriptive and qualitative studies in favour of quantitative, empirical aspects, and designing of techniques of measurement of varied attributes of retailing.

2.1.3 Studies of Proudfoot

In the literature of retailing geography, Malcolm Proudfoot's (1937)[6] pioneering work on Chicago's major outlying business district was perhaps the first to give expression to the form-function analysis. Proudfoot described the geography of retailing in a less complex period than the present when function was relatively single-faceted and related to lines of very limited movement. The strength of the central business district was overwhelming in the concentration of certain kinds of goods in certain types of stores. Outlying districts, therefore, tended to assume clearly defined forms reflecting functions of a lower order hierarchy.

Proudfoot (1937)[7] identified five types of retail structures within the cities of the United States:

(1) Central business district,
(2) Neighbourhood business district,
(3) Principal business thoroughfare,
(4) Outlying business district, and
(5) Isolated store cluster.

This was partly derived from Rolph's (1929)[8] retail district classification which has recognised neighbourhood, string, isolated units and sub-centres as possible retail types of locational forms. Ratcliff (1939)[9] delves more deeply into the matter of store affinity, measurement of the frequency of occurrences of various store types within the retail block in which department variety and chain store etc. were located. In this useful study, buying habits were viewed in relationship to retail structure.

2.1.4 Theoretical Market Place Studies

The theoretical and conceptual study in geography of market centres in the areal context centres round Christaller (1933)[10]. It forms the basis for the building of a hierarchy of central places from the lowest to the highest order based on central goods and services. The identification of this hierarchy and its distribution in space provides for a model framework of spatial organisation when different principles such as marketing, transport and administration functions operate; of these, the first is a market optimising case in which the supply of goods from central place is as near as possible to the places supplied. A higher order central place will serve two of its lower order neighbours. It may do this by serving only two of its six equidistant nearest neighbours and thus having an asymmetric complementary region. This arrangement is termed as K-3 system, i.e., relevant to all studies of retail and wholesale trade service centres.

August Losch (1954)[11] deviating from the fixed K-3 hierarchy but following on similar lines evolved an economic landscape in space, consisting of a lattice of settlement network that presents city-rich and city-poor sectors in his model. However, it is to the Chicago school of the post Second World War period that the world of geographers owe the debt for putting Christaller's central place model on an empirical quantitative base and further research on—

(i) central goods and services as attributes and variates; and
(ii) an operative mechanism for ranking order of settlement to facilitate the building of a spatial hierarchy of settlements.

2.2.1 Post-War II Development of Retail Geography

The Emergency of Quantitative Approach

The post-war period development of retail geography went hand in hand with advance in use of quantitative approach in the field. The retail trading area literature reports several applications of conventional statistical techniques, specifically linear and multiple correlation and regression analysis to market centre studies. This quantitative school presented many studies by Berry (1963)[12], Garner (1966)[13], Rogers (1965)[14], Decay (1964)[15] and Huff (1963)[16]. The quantitative school brought about a shift in approach from the original central place theory which has been well summed up by Scott (1970)[17], "the central place model is static, deterministic and retrospective; it cannot be used to predict the function of shopping centres given the differential growth of multiproduct firms, and the organisational structure of retailing." Hence, the practical utility of central place model became severely limited and hence, the need for newer dynamic notions such as those of Berry and others.

Of those currently engaged in research into retail area structure, probably Huff (1963)[18] has been the most sensitive to the problem of bridging the gap between his statistical formula and the behavioural base of marketing. Huff developed his model on the basis of Reilly's (1931)[19] model of "the Law of Retail Gravitation".

Conventional economic theory ignores the frictional effect of geographical space. However, some attempts have been made to supply the economic model with a spatial dimension. Most notable among such studies are the major works of Hoover (1948)[20] and Isard (1956)[21] taking off from the earlier studies of Losch (1954)[22]. Hoover (1948)[23] is refreshingly empirical in his orientation. His discussion in retail area structure consists principally of reproduction of early trading areas. Isard (1956)[24] discusses retail area structure incidental to his overall consideration of what he terms 'empirical regularities of the space economy'.

2.2.2 Marketing Geography

A meaningful and purposeful product of these studies in marketing in relation to spatial economy is the slow and steady emergence of a distinctive sub-field, i.e., 'marketing geography'. One who has ably guided in defining this field is Applebaum (1954)[25]: "Marketing geography is concerned with the delimitation and measurement of markets and with the channels of distribution through which goods move from producer to consumers". This approach is a sequel to the hiring of geographers by business concerns, trading houses and shopping centres in American cities for spotting location decisions for trade siting.

2.2.3 Decision-making in Retailing

In the realm of theory, the individual reaction to complexity in decision-making has been explored by James G. March and H.A. Simon (1958)[26]. Kevin Lynch (1960)[27] attempted empirically to determine individual reactions to the complexity of the urban environment of three major citities: Boston, Jersey city and Los Angeles.

In his pioneering research effort, he set out to describe and evaluate the character of what he termed the individual's image map, the background against which he argued a person's reaction to a complex urban environment occurs.

2.3.1 Study of Rural Market Centres

Origin of traditional market centres and the pattern of periodic and daily market specially in developing economy and rural societies in Africa, China and India have received considerable early attention. The available literature on retail shopping has been researched by Berry (1967)[28] and his synthesis is the starting point for a geographical study of retail analysis. Berry argues "that the essence of the geography of retail and service business is the clustering of establishments in market centres visited by surrounding consumers. Retail and service business geography is thus inseparable from urban and transport geography. . . central place theory is the theory of the location size, nature and spacing of these clusters of activity and is, therefore, theoretical base of much of urban geography and of the geography of retail and service business."[29] Yet another possible approach in retail market centre is through the study of process that involves the change from the simplest to the most sophisticated form of urban retail establishments. Though location hierarchy and spatial arrangement of retail establishments are not ignored in such studies, they seek to make retailing fully integrated into other processes of economy. While smaller centres of shopping areas locate themselves based upon the decision-making process of disperse individuals, however, in large centres in modern situation such decisions are often made by planner, developer and local authorities.

2.4.1 Studies of Intra-Urban Retailing Hierarchy

Most of the works cited hitherto, by and large, are concerned with the evolution of retailing spatial pattern of individual settlement centres and their organisation areally to develop a settlement hierarchy of nesting orders. Only a handful of these studies relate to retailing in the context of evaluation of intra-urban hierarchy. The study of intra-urban hierarchy in general and functional and retail hierarchy in particular have originated from the broad concept of inter-urban hierarchy. One of the earliest and most comprehensive empirical studies of urban hierarchy ranking of the urban centres of England and Wales was carried out by Arthur E. Smailes (1944).[30] A similar study of urban hierarchy was done by John E. Brush (1953)[31] in South Western Wisconsin in 1953 primarily based on assessment of the business and services existing in centres. Retail and wholesale establishments, banks and financial agencies, trade and personal services, amusement, professional and Government offices were considered in determining each centre's hierarchic status.

2.4.2 Techniques and Methods of Study of Retail Hierarchy

A study of Howard S. Bracey[32] in 1954 in England focused on the area served by a place rather than assessment of goods and services available to which a reference was made by Smailes, six years before 1944 when he discussed the desirability of using quantitative data on bus travel as a meaure of urban rank. Bracey argues that a direct count of shops, services and professions in an urban centre as a method of measuring its centrality is alright for places not exceeding 2000 in population but that for towns and larger villages it is more satisfactory to work from small villages and find out what larger centres they use for what services. As against this approach of Bracey, Mouri Polamaki[33] (1964) in the study of South Bothnian area of Finland measured centrality in quantitative terms simply on the basis of the number of types of central functions occurring in the town. For this, the functions were broken down into separate elements like administrative, wholesale, retail, medical etc. and separately examined for finding the indicator groups and by computing co-efficient of common occurrence. The degree of correlation between the occurrence of functions was attempted to show, according to Polamaki, "the dependence of various institutes on the abstract centrality of central places". Afterwards these results were combined in a partial analysis from which final ranking was derived. B.J.L. Berry and W.L. Garrison (1958)[34] conducted a rigorous test for the ranking of towns in Snohomish country, Washington, in which all the identifiable central place functions, albeit commercial in character, were considered. Subsequently, the techniques of functional classification have been further refined and use of multivariate technique began. Berry used multivariate technique in deriving a combined rank on "aggregate relation and elemental components of central place system" in the early 1960s. Later on the multivariate technique was used by Abiodun (1947)[35] in her work on Nigeria, 'Urban Hierarchy in a Developing Country'.

In contrast to Ms. Abiodun's complex methodology which relies to some extent on subjective decision, Davies[36] formulated a simple but very effective measure—a location co-efficient of a single outlet of any functional type—in his study of South Wales in 1967.

The concept of urban hierarchy based on central place theory was used to rank cities and towns within a region or settlement. But subsequently, the idea of the central place functions was variously carried into diverse fields. It was applied within a city or town based on various functions, often retail function, to find out a central business district with the growth of the city or town. As the commercial function expanded vastly within the centre, it often led to one or more secondary central business districts. In determining the socio-economic status or inequality or well-being within a city also, the concept of urban hierarchy has been used to rank various areas within the city.

W. Burn (1959)[37] in his book *British Shopping Centres* published in 1953 outlined a similar structure which he described as commonly accepted at that time and called it a four-tier system excluding the corner shops.

In this context, it should be mentioned that the concept of hierarchy has been applied for the first time by Carol Hans (1960)[38] to identify intra-city hierarchy of Zurich as against the previously mentioned studies on inter-regional urban hierarchy.

Garrison (1959)[39] and others pointed out that bulk of the existing work on the intra-city retailing was intuitive and based on conventional distinction between convenience shopping and speciality goods.

The problem arising out of a variety of nomenclature in the study of intra-urban hierarchy was solved to a large extent by the summary work of Brian Berry (1962),[40] who drawing on work such as Garrison's has outlined a complete typology of commercial characteristics of American cities outside the Central Business District.

One of the most detailed studies of intra-urban retail nucleation was done by Berry and Garner in 1966 using the area so often worked over in urban studies—Chicago.

B.J. Garner (1966)[41] faced the first problem about how these nucleations were to be defined in areal terms and suggested an objective method based on land value profiles which were constructed along the street and which met at the peak land value intersection.

R.J. Johnston (1966)[42] in an analysis of an intra-metropolitan central place hierarchy in Melbourne related the range of centres of three factors. These were variations in residential density, the socio-economic status of the population and the age of development of an area. In this context, it is worth considering a further study by Ronald Jones (1967)[43] in Edinburgh. A careful method of delimiting centres was employed and a classification derived by a scattergram of total floor space in each centre against the number of selected types of shops.

2.4.3 Application of the Density Gradient Model

It is an established truth that the distance factor plays an important role in explaining the spatial variations from city centre. The application of the density gradient model was first proposed by Belcher (1892)[44] who proved its validity for Frankfurt. In the United States, the demographer Donald Bogue (1949)[45] had investigated the pattern of population distribution around 67 major cities of the United States. Bogue's general conclusion is that the main metropolitan centres dominate the spatial arrangement of population in the United States.

Later, Clark (1951)[46] demonstrated the validity of negative exponential model for density decline based on example for 36 cities. Berry, Simmons and Tennants (1963)[47] further examined the structure of urban population densities for about hundred available examples from various studies drawn from various parts of the world and concluded that regardless of time and place, the model provides a statistically significant fit to the distance of population densities within cities. As the model only provides a statistically significant fit and not the best fit, there were attempts to try variation of exponential models and other models like second degree and density crater function to find out the best fitting models, e.g. by Ajo (1965)[48], Newling (1967)[49], Casetti (1967)[50].

2.4.4 Studies of Intra-Urban Inequalities

An interesting study on intra-urban inequalities was conducted by D.M. Smith (1973)[51]. In his study of Tampa, Florida in 1973, Smith abstracted 47 variables to assess social well-being grouped into six criteria, each broken into sub-criteria and each sub-crietria into variables.

Janet Abu-Laghod's[52] work on Cairo in 1969 is yet another study very frequently mentioned in connection with intra-urban social area analysis. Ms. Abu-Laghod used a set of 13 variables for 216 census tracts for 1947 and 1960. She did a principal component analysis with varimax orthogonal rotation. Some of the variables in her study would not match with western conditions because of the peculiarity of the socio-economic character in a developing country like Egypt.

2.5 Studies of Retail Geography in Urban Setting of India

Studies of marketing geography in general and retail geography in particular in urban setting of India are still on a very limited scale.

2.5.1 *Market Central Place Studies*

Many early studies have been made in different parts of the country in regard to the individual weekly, biweekly and periodical markets. However, studies of the spatial organisation in a network of central places and their hierarchical grouping have been relatively more recent.

The studies of S. Wanamali (1972)[53] L.S. Bhat (1969)[54], S.N.P. Jayaswal (1963)[55], Guha (1964)[56], O.P. Singh (1971)[57], Jayamala Didee (1984)[58], Lahiri (1963)[59] are noteworthy in this regard.

2.5.2 *Studies of Urban Areas and their Functions*

Urban studies have recieved considerable attention among the Indian geographers over the last few decades. The earliest studies were primarily concerned with the history and evolution of individual townships as well as their land use and functional differentiation.

The Madras Geographical Association later renamed the Indian Geographical Journal from Madras and National Geographical Journal published from Varanasi have a large number of contributions in this regard of individual centres. Since the publication of R. L. Singh's (1955)[60] book on Benaras, this field has received much attention from Indian geographers. Following Singh, many Indian cities have been analysed for their urban morphology, functional land use, fringe areas and umlands. Following Jeffersen's (1939)[61] study on the role of umlands in the growth of cities, many Indian geographers have delimited the urban umlands in the Indian context.

2.5.3 *Integrated Urban Analysis in Indian Studies*

While the earlier studies examined the socio-economic and demographic characteristics of urban centres, later studies as for example of Hyderabad by Alam (1965)[62] and of Vijayawada by Prakasa Rao (1975)[63] have revealed a regional multi-disciplinary approach in urban analysis. Studies of models of urban land use and ecology are also recent.

2.5.4 *Spatial Organisation of Urban Centres*

The process of urbanisation and the spatial organisation of urban places in terms of their size and spacing have also received attention especially at the planner's level, with a view of examining the development planning process and the diffusion of economy into the rural area. The contribtions of Pathak (1980)[64], Prakasa Rao (1978)[65] and Moonis Raza (1976)[66] are noteworthy in this regard.

2.6.1 *Studies of Commercial Land-Use*

Though studies of commercial land use with special emphasis on retailing in the urban context are handful, yet there are a number of studies that have contributed in surveying the forces which have helped emergence of larger market centres such as Ramnad (V. Krishnan, 1932)[67], Tinnevelly (V. Krishnan., 1940)[68], Modinagar (A.B. Mukherjee, 1957)[69] and Barpali (N. Pattanaik, 1953)[70].

2.6.2 *Indian Urban Retailing Studies*

Internal structuring of urban market has been studied since 1960. Mention has to be made of the study of Hyderabad by Reddy (1961)[71], Mysore by J.M.C. Beokhalt and P. Vanderveen (1983)[72], Howrah City[73] and Calcutta[74] by Dutt (1966, 1969). In case of Hyderabad, it is observed that location with reference to the potential customer area is decisive in the dispersal of retail vegetable markets, while wholesale vegetable markets had no such connection and emerged as isolated nodes far away from each other.

In the same way, retail system of vegetables in Mysore city has been studied by J.M.C. Beokhalt and P. Vanderveen (1983).

Their paper deals with the intra-urban distribution system of vegetables in the city of Mysore. The study reveals that the existing retail pattern in Mysore city, dominated by a hierarchy of market places, is largely the result of two factors, namely, population distribution and consumer behaviour. Both factors determine the local demand structure, which in turn affects retail behaviour. In vegetable retailing, the informal sector clearly plays a dominant role.

In a study of daily shopping in Howrah city (1966), A.K. Dutt indicated that its pattern was largely determined by community composition, urban functional structure, economic status of the people, and the transport network.

In his Calcutta study, A.K. Dutt (1967) has found out a four level hierarchy of retailing. The norm of central places in Calcutta is 1:4:18:36 which does not conform to any uniform ratio. In this connection, a reference should be made of Berry and Rees' (1969)[75] work on Calcutta. Berry and Rees have also pointed out in their study on Calcutta that "there is increasing functional differentiation of land use alongside regional occupational differentiation."

In the late 1970s some scattered studies have been made in the field of retail geography as well as intra-urban hierarchy by K.M. Kulkarni (1975)[76], Sawant (1978)[77], Sastry (1981)[78], Swaminathan (1979)[79] and Prakasa Rao (1983)[80]. Out of these, Kulkarni's and Sawant's studies were mainly descriptive in nature. Prakasa Rao, Swaminathan and Sastry set out to examine the empirical validity of the step hierarchy vs. continuum in the Indian situation. Prakasa Rao identifies a 4-tier hierarchy for Karnataka at inter-urban level, while Sastry identifies a 4-level hierarchy for the city of Bangalore by applying five different methods. Swaminathan identified a 5-tier hierarchy for Coimbatore by applying factorial analysis.

2.6.3 *Application of Density Gradient Model*

The application of the density gradient model to analyse the density decline from the city centre is empirically well tested in the context of European and American cities.

In the case of Indian cities, the negative exponential model was first used by R.J. Tennant (1961)[81] for Hyderabad and Calcutta. Kar (1962)[82] applied the model to Calcutta and later Brush (1968)[83] applied the model for a number of Indian cities.

Brush examined the gradient pattern of 21 Indian cities of population size varying from 57,000 to 6 million. He fitted the Bleiche-Clark model with slight modification. This generalised application of the model led him to the conclusion that the model does not give a good fit to the density pattern in all Indian cities.

Prakasa Rao (1983)[84] has made such a study of the density gradient pattern of Bangalore city. Two regression models, the first degree exponential and quadratic exponential, were fitted to the density profiles. It was found that the quadratic exponential gave a better fit, though both the models established a statistically significant relationship between population density and distance from the centre. Thus, Prakasa Rao (1983)[85] advocates, "there is a case for re-examining the density pattern in Indian cities based on an analysis with a uniform method of data generation." He further adds, "the fitting gradient models and correlates of density with different types of land-use could be meaningfully used in monitoring the further density structure and pattern of a city."

Some unpublished M.A. dissertations cover studies that include the demographic socio-economic and land use aspects. On the demographic aspects, the pattern of population distribution in Bombay by K. Revati (1971)[86] and occupational structure of the residential areas of Greater Bombay—A spatial review by S.S. Sahasrabudhe (1975)[87] are worth-mentioning. The study on Retail Shopping Function in Fort Area of Bombay by V.M. Adsul (1974)[88] and Masjid Bunder as a wholesale commercial centre by M.C. Pandey (1975)[89] factually describe the functional characteristics of these two commercial zones.

REFERENCES

1. Proudfoot, M. (1937): "City Retail Structure" *Economic Geography*, Vol. 13, pp 425-428.
2. Hoyt, H. (1939): "The Structure and Growth of Residential Neighbourhood in American

Cities", *Reading in Urban Geography* (eds.) Mayer, H.M. and Kohn, C.F., University of Chicago (1967).

3. Christaller, W. (1933): "Die Zentralen Orte in Suddeutschland (Jena)". Translated by W.C. Baskine, 1966—*Central Places in Southern Germany,* Englewood Cliffs, N.J.
4. Smailes, A.E. (1944): "The Urban Hierarchy in England and Wales", *Geography,* Vol. 29, pp. 41-51.
5. Skinner, S.M. (1964): "Marketing and Social Structure in Rural China", *Journal of Asian Studies,* Vol. 24, pp. 3-43.
6. Proudfoot, M. (1937): "Major Outlying Business Centres of Chicago", Private Edition, Chicago University Press.
7. Proudfoot, M. (1937): "Intra-city Business Census Statistics for Philadelphia", U.S. Bureau of the Census, Research Geographer.
8. Rolph, I.K. (1929): "The Location Structure of Retail Trade", Domestic Commerce Series No. 8, United States Bureau of Foreign and Domestic Commerce.
9. Ratcliff, R.U. (1939): "Problem of Retail Site Location", Michigan Business Studies IX, No. 1, Ann Arbor, University of Michigan, cited by Saul B. Cohen and George K. Lewis form and Function in the Geography of Retailing, *Economic Geography,* Vol. 43, No. 1, pp. 1-42.
10. Christaller, W. (1933): *ibid.*
11. Losch, A. (1954): *The Economic of Location* (English Edition, New Haven), translated by William H. Woglam and F. Stolper (New Haven)
12. Berry, B.J.L. (1963): "Commercial Structure and Commercial Blight". Research Paper 85, University of Chicago, Chicago.
13. Garner, B.J. (1966): Internal Structure of Retail Nucleation, North Western University, Evaston.
14. Rogers, A. (1965): "A Stochastic Analysis of the Spatial Clustering of Retail Establishments", *Journal of the American Statistical Association,* Vol. 60, pp. 1094-1103.
15. Decay, A. (1964): "A Probability Model for Central Place Location", *Annals of the Association of American Geographer,* Vol. 56, pp. 550-568.
16. Huff, D.L. (1963): "A Probability Analysis of Shopping Centres, Trading Areas", *Land Economics,* Vol. 53, pp. 81-90.
17. Scott, P. (1970): *Geography and Retailing,* Hutchison, London, p. 16
18. Huff, D.L. (1963): "A Probabilistic Analysis of Consumer Spatial Behaviour", in William S.Dicker (ed.), "Emerging Concept in Marketing", *Proceedings* of the Winter Conference of the American Marketing Association, December, 1962, Chicago, pp. 443-461, cited by D.L. Thompson (1966): "Future Direction in Retail Area Research", *Economic Geography,* Vol. 42, No. 1, pp. 1-43.
19. Reilly, W.J. (1931): *Laws of Retail Gravitation,* Pittsbury Publishers, Inc., New York, 2nd Edition, pp. 19-28.
20. Hoover, Edgar M. (1948): *The Location of Economic Activity,* New York.
21. Isard, Walter, (1956): *Location and Space Economy,* Cambridge.
22. Losch, A. (1954): *ibid.*
23. Hoover, E.M. (1948):*ibid.,* pp. 123-141.
24. Isard, W. (1956): *ibid.,* pp. 55-76.
25. Applebaum, W. (1954): "Marketing Geography", in Preston E. James and Clarence F. Jones (eds.) *Inventory and Prospect, American Geography,* (Syracuse, New York), pp.

245-251.

26. March, James G. and Simon, Herbet, A. (1958): *Organisation*, New York, pp. 136-176.
27. Lynch, K. (1960): *Image of the City*, Cambridge, Mass.
28. Berry, B.J.L. (1967): *Geography of Market Centres and Retail Distribution*, Prentice Hall, Englewood Cliffs.
29. Berry, B.J.L. (1967); *ibid.*, p. 3.
30. Smailes, A.E. (1944): *ibid.*
31. Brush, J.E.(1953): "The Hierarchy of Central Places in South Western Wisconsin", *Geographic Review*, XL. pp. 380-402.
32. Bracey, H.E. (1954): "Towns as Rural Service Centre, An Index of Centrality with Special Reference to Somersit", Transactions of the Institute of British Geographers, Publication No. 19, pp. 95-105.
33. Polamaki, M. (1964): *The Functional Centres and Areas of South Bothnia*, Finland, Fennia, 88-1.
34. Berry, B.J.L. and Garrison, W.L. (1958): "The Functional Bases of Central Place Hierarchy", *Economic Geography*, Vol. 34, pp. 145-154.
35. Abiodun, J. (1947): "Urban Hierarchy in a Developing Country", *Economic Geography*, Vol. 43, pp. 347-367.
36. Davies, W.K. (1967): "Centrality and the Central Place Hierarchy", *Urban Studies*, Vol. 4, pp. 61-79.
37. Burn, W. (1959): *British Shopping Centre*, London.
38. Carol, H. (1960): "The Hierarchy of Central Functions within the City", *Annals of the Association of American Geographers*, Vol. 50, pp. 419-438.
39. Garrison et al., (1959): "Emperical Verification of Concept of Spatial Structure", Studies of Highway Development and Geographical Change, University of Washington Press, pp. 67-99.
40. Berry, B.J.L. (1962): "The Commercial Structure of American Cities: A Review", Community Renewal Programme, Chicago.
41. Garner, B.J. (1966): The Internal Structure of Retail Nucleations, North Western University, Evanston.
42. Johnston, R. J. (1966): The Distribution of an Intra-metropolitan Central Place Hierarchy in Melbourne", *Australian Geographical Studies*, Vol. 4, pp. 17-33.
43. Jones, R. (1967): "Central Place Theory and the Hierarchy and Location of Shopping Centres in a City: Edinburgh", cited in Carter, H. (1972): *The Study of Urban Geography*, London, p. 119.
44. Belcher, M. (1892): "Statistiche Beschreib, Derstadt, Frankfurt, Main UndThrer, Bevole-Kerung", Frankfruit Am main.
45. Donald, Bogue (1949): "The structure of Metropolitan Community, A study in Dominance and Sub-Dominance", Ann Arbor at Michigan, cited by V.L.S. Prakasa Rao (1983), *Urbanisation in India*, Concept Publishing Company, New Delhi, p. 197.
46. Clark, C. (1951): "Urban Population Densities", *Journal of the Royal Statistical Society*, Sec. A.C. XIV, Part IV, pp. 490-496.
47. Berry, B.J.L., Simmons, J.W., and Tennants, R. J. (1963): "Urban Population Densities, Structure and Change", *Geographical Review*, Vol. 53, pp. 389-405.
48. Ajo, R. (1965): "On the Structure of Population Density in London", Field Acta Geographica 18, cited by Prakasa Rao (1983): *Urbanisation in India*, Concept Publishing Co., New Delhi, p. 44.

49. Newling, B.E. (1967): "The Spatial Variations of Urban Population Densities", *Geographical Review*, Vol. 59, pp. 242-265.
50. Casetti, E. (1967): "Urban Population Density Pattern, An Alternative Explanation", *Canadian Geographer*, 11, pp. 25-49.
51. Smith, D.M. (1973): *Geography of Social Well-being in the United States*, New York, Chapter, 9, pp. 120-134.
52. Abu-Laghod, J.L. (1969): "Testing the Theory of Social Area Analysis; A case of Cairo, Egypt", *American Sociological Review*, Vol. 34, No. 2, pp. 198-212.
53. Wanamali, S. (1972): "Central Place and their Tributary Population: Some Observations", *Journal of Behavioural Sciences and Community Development*, Vol. VI, No. 1, pp. 11-39.
54. Bhat, L.S. (1969): "Central Place Model as a spatial Framework for Regional and National Planning in India", Symposium on the City as a Centre of Change in Asia, University of Hong Kong.

    Bhat, L.S. and others (1974): "Central Place Hierarchy for a Developing Agricultural Region: A Case Study of Karnal Area". Report of a Study sponsored by ICSSR, New Delhi.
55. Jayaswal, S.N.P. (1963): "Bindki: A Study in the Evolution of a Trade Centre", *National Geographical Journal of India*, Vol. 9, pp. 187-200.
56. Guha, B. (1964): "Morphology of Burdwan", The Earth.
57. Singh, O.P. (1971): "Towards Determining Hierarchy of Service Centres" A Methodology for Central Place Studies", *National Geographical Journal*, Vol. 17, pp. 90-105.
58. Jayamala, Didee, (1984): *Central Places of Western India*, Pune.
59. Lahiri, T.B. (1963): "Calcutta Metropolitan District", *Proceedings* of Indian Science Congress.
60. Singh, R. L. (1955-56): "Urban Hierarchy in the Umland of Benaras", *Journal of the Scientific Research*, Banaras Hindu University.

    Singh, R.L. (1955) *Banaras: A Study in Urban Geography*, Nand Kishore and Bros., Benaras.
61. Jefferson, Mark (1939): "The Law of the Primate City", *Geographical Review*, Vol. 29, pp. 226-232.
62. Alam, S.M. (1965): *Hyderabad-Secunderabad Study in Urban Geography*, Allied Publishers, Bombay.
63. Prakasa Rao, V.L.S. and Ramchandran, R. (1975): *Socio-Economic Survey of Vijayawada City*, Mimeographed, ICSSR, New Delhi.
64. Pathak, C.R. (1980): "Urbanisation Problem and Policies in India", *Indian Journal of Regional Sciences*, Vol. XII, pp. 71-90.
65. Prakasa Rao, V.L.S. and V.K. Tewari, (1978): *Urbanisation in India—Spatial Dimensions*, Mahadev, P.D., (ed.), Urban Geography, Contribution to Indian Geography, VII, pp. 193-214.
66. Raza Moonis, and Habeeb, A. (1976): "Characteristics of Colonial Urbanisation—A Case Study of the Satellite 'Primacy' of Calcutta (1850-1921)", in Alam Pkskirhwrky (ed.), *Urbanisation in Developing Countries*, pp. 187-218.
67. Krishnan, V. (1932): "Fairs and Trade Centre of Madurai and Ramnad District", *Journal of Madras Geographical Association*, 7, No. 3, pp. 236-249.
68. Krishnan, V. (1940): "Trade Centre of Tinnevelley District", *Journal of Madras Geographical Association*, 15, No. 3, pp. 272-276.
69. Mukherjee, A.B. (1957): "The Bi-weekly Market at Modinagar, *Indian Geographical*

Journal, Vol. 32, pp. 35-42.

70. Pattanaik, N. (1953): "Weekly Market at Barpali", *Geographical Review of India,* 15, No. 1, pp. 151-171.
71. Reddy, S.N. (1961): "Vegetable Market and Regional Relationships of Hyderabad City", *Geographical Review of India,* 23, No. 3, pp. 24-40.
72. Beokhalt, J.M.C. and Vanderveen, P. (1983): "Retail System of Vegetable in Mysore City, India", Mahadev, P.D. (ed.), Urban Geography, Contribution to Indian Geography, VI, New Delhi, pp. 110-131.
73. Dutt, A.K. (1966): "Daily Shopping in Howrah", *National Geographical Journal of India,* 12, No. 3, pp. 163-167.
74. Dutt, A.K. (1969): "Intercity Hierarchy of Central Places, Calcutta—A Case Study", *Professional Geographer,* Vol. 21, pp. 18-23.

    Dutt, A.K. (1967): "Shopping in Calcutta", *Town Planning Review of India",* Vol. 37, pp. 207-216.
75. Berry, B.J.L. and Rees, P.H. (1969): "The Factoral Ecology of Calcutta", *American Journal of Sociology,* Vol. 74, pp. 448-491.
76. Kulkarni, K.M. (1975): "Shopping Patterns Near Greater Bombay—Suburban Railway Station—A Case Study", *Bombay Geographical Magazine,* Vol. XXIII, No. 1, pp. 10-25.
77. Sawant, S.B. (1978): "Shopping Area in the City of Poona", *Geographical Review of India,* Vol. 40, p. 155.
78. Sastry, G.S. (1981): "Retail Trade Structure of an Indian Metropolis-Bangalore", *Annals of the National Association of Geographers, India,* Vol. 7, No. 2, pp. 57-73.
79. Swaminathan, M.S. (1979): "A Qualitative Appreciation in the Study of the Spatial Distribution and Functional Hierarchy of Shopping Centre in Coimbatore City", *National Geographical Journal of India,* Vol. 25, pp. 231-254.
80. Prakasa Rao, V.L.S. (1983): *Urbanisation in India,* Concept Publishing Co., New Delhi, p.68.
81. Tennant, R.J. (1961): "Population Density Pattern in Eight Asian Cities", paper read at West Lakes Division meeting of the Association of Amercian Geographers.
82. Kar, N.R. (1962): "Growth Distribution and Dynamics of the Population Load in Calcutta (Mimeo), cited in V.L.S. Prakasa Rao, *Urbanisation in India,* Concept Publishing Co., New Delhi, p. 97.
83. Brush, J.E. (1968): "Spatial Pattern of Population in Indian Cities", *Geographical Review,* Vol. LVIII, July, 1968, p. 374.
84. Prakasa Rao (1983): *ibid,* p. 211.
85. *Ibid,* p. 212.
86. Revati, K. (1971): "Pattern of Population Distribution in Bombay", Unpublished M.A. Dissertation.
87. Sahasrabudhe, S.S. (1975): "Occupational Structure of the Residential Areas of Greater Bombay—A Spatial Review", Unpublished M.A. Dissertation, Department of Geography, University of Bombay.
88. Adsul, V.M. (1974): "Retail Shopping Function in Fort Area of Bombay", Unpublished M.A. Dissertation, Department of Geography, University of Bombay.
89. Pandey, M.C. (1975): "Masjid Bunder as a wholesale Commercial Centre", Unpublished M.A. Dissertation, Department of Geography, University of Bombay.

3

EVOLUTION AND GROWTH OF RETAILING TRADE IN BOMBAY
A General Survey

3.1 Pre-colonial Bombay

The main city of Bombay is built over a group of seven islands that were known as "Heptanesia"[1] to the western world at the beginning of the Christian era. The real growth of Bombay to its present status as the premier port of Western India is a sequel to the colonial trade interests and settlement initially of the Portuguese in the late 16th century, followed by the transfer of trade interests and property to the British King in 1661 as a part of a marriage treaty of a British King with a Portuguese princess.

What pre-existed the colonial settlement was a group of fishing and farming villages spread over the seven islands and the medieval Hindu township of Mahim on its decline at the north end of the group of isles. In all probability, this primary producer-population living in rural communities in the seven islands did not have any significant exchange economy. Little is known and recorded about the local consumer trading of these early days. No doubt, Mahim lying at the head of the Mahim Bay had a commercial "bandar" with significant coasting and overseas trade, as recorded in medieval records of history and navigation.[2]

The only pertinent reference to the local trade characteristics of Mahim refer to A.D. 1677 when a mention is made of the presence of the petty custom house of the chief settlement that had a "reasonably handsome bazaar" at one end "looking into the field, where cows and buffaloes graze".[3]

3.2 The Colonial Era: Rise of Overseas Commerce

The colonial settlement of the Portuguese, built round the Quinta or the Manor house in the 16th century, was later on improved by the British East India Company in the 18th century to develop a fairly well-fortified town, comprising within it the Manor house, the Mint and the Custom House, all adjoining the Pier and landing point at the south end. "The foundation of English dominion in Bombay lies in the 11th article of the marriage treaty concluded on 23rd June 1661 between his Majesty Charles II, the King of Britain and Alfonsus VI, King of Portugal", writes James Douglas.[4] The English had to wait till 1665 for the actual possession of Bombay, when Humphry Cooke personally took possession of Bombay on February 18 of the same year.[5] Cooke was himself invited by Indian traders to settle in Bombay to carry on business, thus quickly making a beginning of the new era of Bombay which flourished to a great extent in subsequent years.

However, the consolidation of trade and commerce of Bombay had to await the leadership of Gerald Aungier who governed this island from 1669 to 1677. Aungier laid the basis of the "administrative, commercial and social expansion" of Bombay.[6]

Before the arrival of the British, trade in Bombay was confined to cocoanuts, dry fish and coir, but by 1670 the island's trade expanded in a big way. Bombay's local trade was composed of bullion, wine, tobacco, opium, cocoanut and coir, and its chief imports from England were clothes, copper, lead, silver, iron and iron-wares.[7]

Trade and commerce picked up when Charles Boone became Governor of Bombay. He fortified the city from outside attack and built a wall around it. The strengthening of Bombay's defence coupled with the decline of the Sidis and the Marathas on the mainland led to the "local bankers from Aurangabad and Poona to open business in Bombay, attracted by the fact that Bombay had become a central point between Western and Upper India on the one hand and Malabar Coast, Persian Gulf and the Red Sea on the other".[8]

By the close of the third decade of the 18th century Bombay's trade and commerce was in a most flourishing condition. A large volume of cotton imported by boat from Gujarat ports was transferred at Bombay into larger sailing vessels which then sailed to Madras, Bengal and China.

3.2.1 *Wholesaling Commerce*

The first local banking interests came into existence around 1750. The cotton trade with China commenced around 1770. Till early 19th century, trade was mainly in the hands of Indians and a couple of British agencies apart from the East India Company. The primary product of trade till the middle of the 19th century was cotton which used to be brought from up-country by boats and country vessels, and unloaded near the initial harbour behind the factory. The open yard of Cotton Green located in the area called Horniman Circle was the most striking commercial landscape feature of the original fortified European town.

Though the Cotton Green was shifted subsequently to Colaba and still later to the warehouses of the present day Cotton Green in the north of the city, it is the Cotton Green and the adjunct of the warehouse and harbour yards behind original Bombay Castle that was the original wholesaling and commercial nucleus of Bombay, well within the original fortified Eurpoean town.[9]

3.2.2 *Seeds of Retail Trade in the European Town*

Although the wholesale trade in cotton primarily and in other commodities like grain and edible oils was the basis of the commerce of Bombay, however to sustain the population living within the Fort, it was necessary to develop some form of retailing trade. Here developed the embryo of the retail trade function of Bombay. The native population living outside the Fort to its northwest in Girgaum, Cavel and Chowpatty areas could offer rice, cocoanut, fish and perishables in the town market. As early as 1688, a community of weavers from Chaul had been brought to settle on a freshly built street stretching from the Custom House to the Court. A number of Surat traders belonging to the bania community were also encouraged to migrate and settle down in Bombay and were assured of trade privileges. So too, the Parsees were also encouraged to settle within the fortified town.[10] This promotion of men of sound commercial knowledge was with the intent of expanding and promoting local and external trading. The trade area developed on the right of the Churchgate on way to the native town but within the fortified town. The bazaar "to the south of Church Street having become a great nuisance"[11] had three north to south running streets, namely, the Bora Bazaar Street, the Bazaar Gate Street and the Frere Road that constituted the main retail trading zone of the fortified European township. The native traders from the "black" town entered through the Bazaar Gate and sold their commodities and returned home late in the evening. In the words of Gerald Aungier, "All provisions and sustenance necessary for life are procurable at Bombay; all sorts of corne, and graine for man and beast: there is beef, mutton, veal, lamb, pork, hens, ducks, geese, fish, very good in their kind and sufficient quantity, but not all the produce of the island itself the greatest part being brought from the neighbouring main land."[12]

The first local market in the fortified town appears to have been established for sale of fruits, vegetables and flowers in A.D. 1696 and dealers in these commodities were permitted to occupy the market, free of rent.

In 1769, two large sheds were built for the public market to be held, one for meat and fowls and the other for fruits and greens. During the administration of Mr. Hornby (1776-84), a market had been constructed which was destroyed in 1803 by a great fire. This market was rebuilt during the administration of Duncan. In the enclosure of Mody Street, a market popularly called "breakfast market" (shown in Map 3.1) was constructed in 1847 mainly for supplying bread, eggs and breakfast materials to the local Parsi and other rich Indian communities.

3.3 Early Physical Growth of Bombay

The first half of the 18th century witnessed an active expansion and promotion of commerce of Bombay accompanied by rapid growth of population and improved sanitation condition. The process of reclamation had commenced the process of welding together of individual islands by bridging the breaches in between and infilling of the inland seas and lagoons. A number of roads on embankments were built on reclaimed ground as the forerunner to develop built-up area. A variety of native communities, artisans, craftsmen and traders were actively encouraged to settle on the newly reclaimed areas outside the old town.

Builders were invited to migrate from Surat to Bombay and to set up the Bombay dockyard. The Bazaar Gate became a continued concourse of people to and from the town. The first local banking interest came into existence around A.D. 1750 and the cotton trade with China commenced around A.D. 1770.

A description of Bombay given in the Oriental Memoirs (1775) states: "The large bazaar or the street in the black town within the fortress, contained many good Asiatic houses and shops filled with merchandise from all parts of the world, for the Europeans and natives. These shops are generally kept by the Indians, especially the Parsis, who after paying the established import customs were exempted from other duties."[13]

The great fire of 1803 caused a heavy destruction of the bazaar within the black town. Apart from the direct destruction from fire, many buildings had to be pulled down on grounds of safety. However, the fire

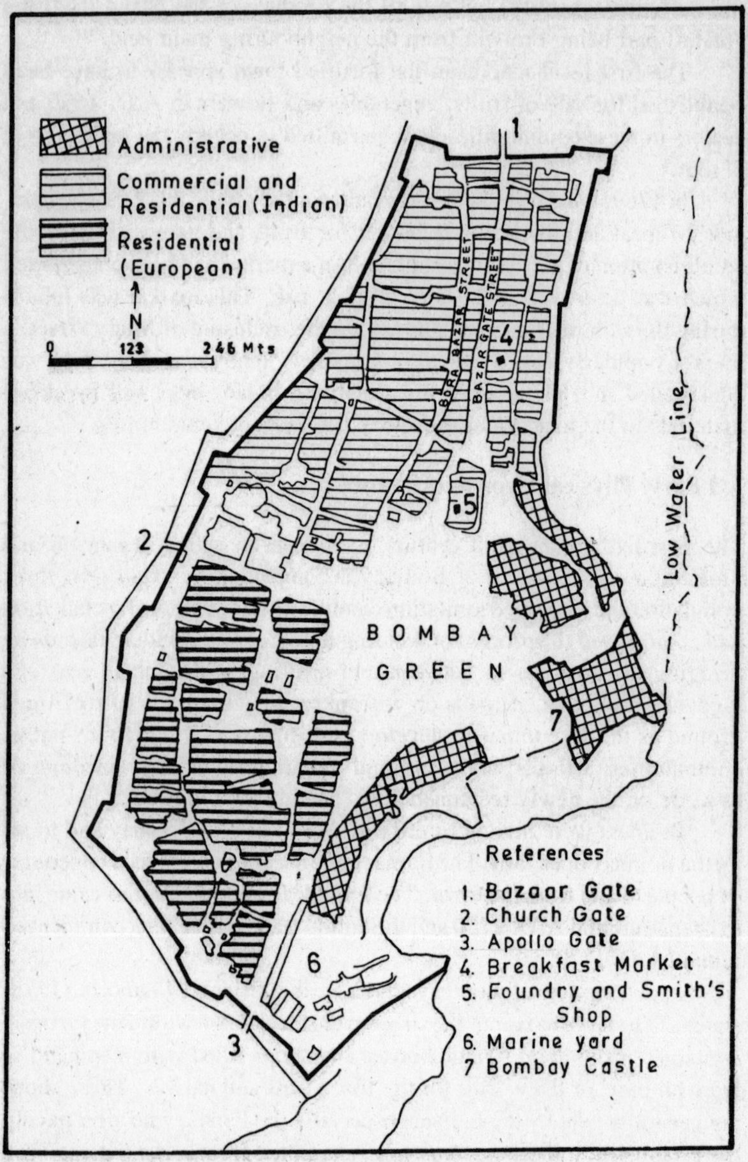

Map 3.1 Commercial Land-use of Fort Area in Early Colonial Period

of 1803 turned out to be a blessing in disguise. Its importance lay not so much in the improvement of the town within the fort wall but the inducement it offered to the construction of a new town outside. The area north of the Fort had only been sparsely built up till then and it was considered by the town committee,

> "to convince the natives in question of unadvisableness of their residing in a garrison crowded with lofty structure filled with goods and merchandise and intersected by such narrow streets as existed before the fire."[14]

The original bazaar accommodation within the Fort was to be left to the more respectable and wealthy merchants. Trading communities were to be settled in the localities outside the fortification but near to the docks.

3.3.1 *Development of the "Market" outside the Fort*

From 1803 onwards, traders' residence, godowns, shops and markets began to rise in the area that today forms the nucleus of Mandvi, Chakala and Umarkhadi. By 1860, eight families of Jains were settled in what is now the Dongri street. The succeeding years saw fresh arrivals of Kutchi tradesmen who settled in that area. The Khojas who were the hawkers of parched rice and the Moors from the Mahim area settled around the four main bandars, namely, Bori-Bandar, Kolibandar, Masjid-Bandar and Chinch Bandar, to mould the area of Mandvi and Masjid. The foundations of the commercial nucleus of Bombay were thus laid in these years just outside the fortified town and its garrison, Fort St. George. It was separated by the Green of the Esplanade from the European town and was equally accessible to the original native town of Girgaum-Cavel and the European town within the Fort and also close to the newer developments of the docks that were still in the throes of development.[15]

3.3.2 *Retail Markets of the 19th Century*

By the middle of the 19th century, the town was spilling over beyond the fortified European town and the old native town, further north into the newly reclaimed areas and the peripheral areas, the latter no doubt being preferred by the Europeans because of the better location and hygiene.

Mrs. M. Postan wrote in 1839 about the fort thus:

"The Fort contains three of those attractive places known as ruination shops: they are filled with a very tolerable collection of haberdashery which is replenished on the arrival of any French or English ships."

"In addition to the mercers, the Fort contains two good chymists, two booksellers, a musical, circulating library, and an English Jeweller."

"Within the Fort, are two large bazaars, the *China bazaar*, as it is called, being one and the *Thieves bazaar,* the other. The first contains all sorts of curiosities with storeroom full of China satins and tea-boxes, the second is crowded with warehouses, filled with European articles, to be disposed off at a small profit."[16]

At that time, the native town contained "three principal bazaars",[17] which were linked through numerous serpentine cross-roads. These markets were rather dirty and thè Europeans and the Indian Gentry patronized these markets generally on Sunday when business reached its peak. The markets were filled with goods in the morning and by noon were crowded with a variety of people, viz., "Arabs, Persians, Jains, Fakirs, Padres, Jews and Parsis". The Arabs used to deal in horses and had their stables occupying a considerable space in the bazaar. The most profitable trade carried on in the bazaars, however, was the sale of toddy; and even after the spacing of toddy shops were regulated, about "every sixth shop advertises its sale".[18] Another quotation from Mrs. M. Postan (1835) would provide a pen picture of the bazaars of Bombay during the middle of the 19th century:

"The Bazaars are lined with open shops, in which all descriptions of workmen are to be seen, occupied in their various callings. To a stranger's eye, the Chintz bazaar (presumably the Tamarind Market or the Tamarind Street) will afford the most curious scene; the road skirts that particular part of the bay occupied by native shopping and its wholly devoted to the purposes of commerce. Here indeed is a "mart of nations", where the genius of traffic reigns triumphant, and the merchandise and produce of all nations of the East seem garnered in one common store, awaiting an escort to the lands, where the arts and manufactures of civilised life will increase the value of Nature's gifts. Piles of rich gums and aromatic spices, carboys of oil and rose water, pure ivory from the forests of Ceylon, rhinoceros hides from the burning coasts of Zanzibar, the richest produce of Africa, India, Persia and Arabia is here cast in large

> heaps, mingling with coir, cables, huge blocks and ponderous anchors, the requisite material of island exportation."[19]

The commercial nucleus of the native town beyond the Fort walls was a sort of an extension of the north of Fort; it was a commercial replica of the European town of Bombay. Writing almost nine decades later than Postan, Falkland remarks that the basic character of the Indian market has not changed much but the market did extend further towards the north into areas such as the present-day Bhendi-Bazaar area while commercial activities in the core area became more intense.

> "The streets from that part of the bazaar which is called the Bendy Bazaar to the Esplanade is crowded from sunrise to nine O'clock at night."
>
> "The most interesting part of the native town begins at the Bori-bazaar where in the cool of the evening, the Persian and Arab horse dealers sit in open air sipping coffee."
>
> "In the East it is usual for all the members of a trade to line in the same vicinity and a row of many shops tenanted by copper-smiths then by cutters of stones, by vendors of gold and silver ornaments, of wearing apparel for the natives each having their proper locale. There are sellers of flowers for weddings and offerings at temples, shops where rice, split-peas, salt, oil, vinegar, ghee, butter, betelnut, paan leaf and fruits are retailed besides confectionary dealer in and tobacco or copper vessels for household use among the natives."[20]

By about 1850, apart from the two retail bazaars in the European town within the Fort, three other bazaars—the *China bazaar* of Girgaum, Kalbadevi-Bhuleshwar, and Mandvi-Masjid-flourished meeting the consumer needs of the growing population of the city.

3.3.3 *The Commercial Landscape of 19th Century*

An essential character of the commercial function, more so of retailing, as it developed during the 19th century is its fascinating and colourful variations. Most of the Indian traders during the middle of the 18th century were Banias, Muslims, Parsis, Jains, Bohras, Memons, Jews and of course, Europeans. The Parsis, active, industrious and clever, were ranked only next to the Europeans. Elwood[21] mentioned in 1830 that the retail trade of Bombay was almost wholly in the hands of this enterprising community, who managed to come successfully through the commercial depression of 1827 and 1834, and even outlived some of the European mercantile firms as well.

Caste and commodity-wise segregation started during this period which to the day continues to be an important feature of Bombay's trade. Grain trade was dominated by Gujarati Banias (Kutchi), and to some extent Lohars. Retail cloth shops were owned by Bhatias and Banias and also by some Muslims. The wholesale business of cloth was dominated by Muljis, Jethas and other big cloth merchants. The fruits and vegetable trading was largely dominated by the Marathas coming from Deccan in some cases, Banias coming from Kutch, Khojas and Memons.[22]

3.4 The Bombay Industrial Era

The industrial era of Bombay started with the introduction of railway from Bombay to Thana in 1853 and its subsequent extension in the next two decades to directly link the port city of Bombay with the agriculturally rich and productive hinterland that feeds Bombay's trade and commerce. The Great Indian Peninsular (GIP) Railway, and Bombay Baroda (BB) and Central Indian (CI) Railways helped in carrying large volumes of commodity traffic to and from Bombay. With the tremendous growth of foreign trade carried from Bombay port following the opening of the Suez Canal in 1867, cargo movement of railway also increased.

On the industrial front, the first textile mill was founded in Bombay in 1854 and was soon followed in the succeeding decades by the coming up of a large number of new mills, all located beyond and north of the native town and commercial area, on newly reclaimed ground that was then peripheral to the city but which now lies in the heart of the city in the area stretching from Byculla in the south to Dadar in the north. The industry at the start was biased heavily in favour of spinning to cater to the needs of the Chinese foreign markets.

3.4.1 *Further Physical Growth*

The city was gradually creeping over the reclaimed higher ground westward along the Backbay and northward to Byculla. The original fort walls, a mere encumbrance, were done away with. The cotton yard was shifted further south to Colaba to be close to the landing point near Apollo Bandar.

3.4.2 *New Commercial Interests*

The residential area in the southern Fort area shifted further south to Colaba to give place to the growing needs of administration and business and simultaneously a new retail shopping area started taking shape on the newly developed Colaba Causeway.

During this period, cotton trade generated intense commercial activity. After a short-lived business boom came a big bust during 1861-1866, historically known as the *Share Mania.*[23] The enormous wealth generated from the export of Indian cotton to Great Britain following short supply of American cotton in British markets led to intense speculation which helped the mushroom growth of banks and financial houses during this period just to collapse almost overnight when the business cycle turned down-trend.

Intra-city land transport network also expanded to meet the growing needs. The tramways introduced in 1872 was an indicator that the urban system was becoming mature. The number of textile mills increased by seven times from 10 to 70 between 1870 and 1890. These mills were generally located in the central part of the island, viz., Parel, Lalbaug area employing a large number of people who came from Bombay's immediate catchment areas in Maharashtra. This growth of the early textile mills of Bombay was not only a prelude to the tempo of industrialisation but also the rapid urbanisation of Bombay, and enormous growth of population numbers mainly through immigration.

The opening of the Suez Canal in 1867 shortened the distance to British and Western European ports and brought about a rapid expansion in the port trade activity. A number of docks were built during the second half of the 19th century to cope with the expanding volume of sea-trade.

Trade continued to flourish hand in hand with industrial activities, though manufacturing was gradually becoming more important. This is reflected in the high growth of labour force in cotton textile industries. from less than 7000 workers in 1860, workers in cotton textile mills increased by over four times within 20 years to 30,000 in 1880 and further by 2.5 times to 73,000 in 1900 and further to 1,70,000 in 1981.[24]

3.5 The 20th Century

The turn of the 20th century saw the collapse of the Chinese cotton market. At the same time, the rise of the *Swadeshi* movement came to the rescue of the cotton mill industry which made a vital shift from spinning to weaving. This necessitated a change in market from overseas to inland and local ones.

3.5.1 *Industrial Diversity*

Post-World War I period brought about the first symptom of diversification of the industrial base of Bombay. New areas of industrial activity, viz., chemicals and oil mills, structural machines and small engineering units, came on the scene.

This industrial growth in Bombay was further facilitated during the post-independence period with the Government policy and sizeable industrial inputs. As a result, a wide range of light and medium engineering, chemical and drug industries, refineries and associated petrochemicals and fertilizers not only came into being but also started prospering with increased number of persons employed in them. This in turn helped the growth of trade and commercial activities, both wholesaling and retailing.

3.5.2 *Demographic Growth*

The rapid growth of the industrial, port and trade functions of the city since early 19th century brought in its wake enormous increased in population. The old city area—European and native—taken together could not accommodate all the increase in residential population as well as the newer urban functions. This inevitably led to the growth of an urban sprawl which quickly submerged the rural fringe. The vast expansion in industrial activities further increased the commercial and port functions of the city. The entire eastern water-front extending from Colaba to Trombay started humming with port activities, many business houses and financial institutions came into existence and warehouse and bulk handling facilities increased. A wholesale trade area emerged in the area to the north of the Fort, south of the old Hindu and Muslim towns and adjoining the docks close to the V.T. Railway terminal in the market areas. Simultaneously, the cotton yard was shifted north to its present location of Cotton Green, closer to the mill area. Soon the city engulfed the whole island and spread further beyond into Salsette, developing suburbs and absorbing them all into a well-knit city of a metropolitan dimension.[25]

3.5.3 *The Urban Sprawl*

The original urban nucleus being eccentrically located in the south of insular Bombay, all subsequent horizontal expansion was inevitably northward, to engulf the entire island area by 1940, and for the subsequent spillover to spread further north along the rail-cum-road corridor in suburban Salsette in the post-fifties. The retailing nucleaus of South Bombay was no more readily accessible by the thirties of the 20th century to the population of North Bombay, necessitating an inevitable dispersal.

3.5.4 *Fresh Trends in Retailing*

In 1839, there were in fact two important shopping enclaves within the Fort area, and native town "contained three principal bazaars"[26] which was about 3 km away from the Fort. From 1839 to 1930, from Mrs. Postan to Mrs. Falkland, it appears that the character of the Indian bazaar remained the same except for its tendency for northward expansion, along the fringes and a gradual permeation along the main roads into the older residential areas like Girgaum Road, Charni Road and Opera House.

From 1930 to 1985, the total retailing landscape has undergone a total transformation. The retail function which was restricted upto 3 to 5 kms from the original nucleus of the city till 1930, today extends upto 40 kms from the same old nucleus throught the eastern and western commuting corridors, and from them on either side along trunk feeder roads, with a number of alternative retail cluster catering to different residential areas nearby. Naturally, higher and lower order retail centres have developed and definitely there are qualitative differences among various centres, but a well-knit retail establishment pattrern has emerged within the urban system of metropolitan Bombay.

An overall and gross qualitative assessment of the intensity of this metropolitan dimension can be made from the nature of rapid expansion of retail establishments. The city had 3,957 shops in 1890 according to the Gazetteer of Bombay; [27] this number increased to 1,35,545 in Greater Bombay in 1983 according to the BMC record, thus reflecting a growth of 34 times in 153 years. This is not only perhaps the largest in number in any Indian city but one of the highest in any South-East Asian city. This gigantic growth is not only in number but also in variety. Retailing has also acquired a distinct personality in terms of spatial distribution.

Since the thirties of this century, suburban development in the north of the island and in areas around Dadar, Mahim, Matunga, Sion and Wadala was on a brisk pace, following the electrification of the suburban railways. Town planning schemes like the Shivaji Park Scheme (1928), the Matunga-King Circle Scheme (1940), the Sion-Wadala Estate (1950-70) were put into effect, the newly built-up residential areas developing retailing complexes in the neighbourhood itself as a part of the plan, at the rail head and along the main arterial functions, like the Khodadad Circle (Dadar T.T.), Maheswari Udyan (King Circle) and Rani Laxmibai

Chowk (Sion Circle). By early fifties, the city island was nearing saturation of built-up areas.

The post-fifties saw a spurt in industrial activity in the city. The municipal limits of the city Corporation were twice enhanced in quick succession in the fifties (i.e., 1951 and 1956) to bring within the urban orbit rural fringes that could be suburbanised to take in fresh waves of population influxes into the city, and the excess spill-over from the city itself. Newer industrial growth invaded the suburbs, especially along the central rail corridor, the Powai industrial estate and Saki-Marol area on the Andheri-Kurla cross link and much more recently the outer western rail suburbs beyond Andheri. Restricted industries developed around Deonar-Trombay. What initially started as a string of dormitories to provide shelter to the commuting city workers themselves started developing secondary and tertiary functions. The newer areas of suburban development started acquiring retailing functions, of course with a preference in site at the rail heads, or on roads leading to them, taking advantage of convergence of the commuting population inflow and outflow. Thus, a new set of retail clusters started emerging in the suburban areas in the sixties, seventies and eighties, primarily to cater to the local and nearby residential population, although over the years they have also developed diversification and specialisation, as will be seen in later chapters.

3.6 Broad Summary

From the foregoing review, it will be thus seen:

(i) The initial commercial nucleus of colonial Bombay was within the fortified European town, primarily to cater to the wholesale sea-trade in the raw materials export, especially of cotton, but developing as an adjunct a limited retail function in the north of the Fort, to cater to the day-to-day needs of the local resident population.

(ii) Native trading communities with specific interest in particular commodities were encouraged to settle in the city and were extended particular privileges and concessions with the intent of promoting trade.

(iii) The first major physical outgrowth of commercial activity took place after the great fire of 1803, when the commercial cluster of Kalbadevi-Mandvi-Masjid was developed outside the native and European residential area, but close to them and the docks. This area over the years has intensified and diversified its activity but with strong social affinities in-built, arising out of long traditional associations of particular trade activities.

(iv) Till 1930, the further expansion of trade, especially retail trade, was by way of peripheral growth from the nucleus area with progressive intrusion along the main roads into the residential areas and the emergence of a new retail shopping area south of the Fort, in Colaba Causeway.

(v) A further stage of retail activity expansion came in the forties and fifties of this century with the rise of new residential areas in the north of city island; this brought in its wake some retailing clusters in the north of city, while more recently commuting rail head-cluster of retsilaing in many suburbs have emerged. There is thus a progressive dispersal and decentralisation of retialing all over the metropolis gradually replacing the earlier centralised shopping in the down-town, south.

REFERENCES

1. Edwards, S.M. (1902): *The Rise of Bombay: A Retrospect,* Times of India Press, Bombay, p.3.
2. Edwards, S.M. (1902): *idid,* Chapter 1, pp. 1-47.
3. Edwards, S.M. (1902): *ibid,* p. 198.
4. Douglas, J. (1893): *Bombay and Western India: A Series of Stray Papers,* Vol. 1, London, p. 37.
5. Edwards, S.M.(1902): *ibid.,* p. 37.
6. Anderson, P. (1856): "The English in Western India", pp 124-129, *Cited in Gazetteer of Bombay,* Vol. 1, p. 25.
7. *Gazetteer of Bombay City and Island,* (1909), Vol. II, Bombay, pp. 67-68.
8. "Bombay Government to Court of Director in 1753", cited in Warden, F. (1861): "Report on the Landed Tenure of Bombay 1814, Bombay", quoted from *Gazetteer of Bombay City and Island,* Vol. I, p.119.
9. Graham, M. (1812): "Journal of a Residence in India", Edinburg, p. 13, cited in *Gazetteer of Bombay City and Island,* (1909), Vol. III, pp. 43, 45, 54.
10. *Gazetteer of Bombay City and Island,* (1909), Vol. III, p. 66.
11. Edwards, S.M. (1902): *ibid.,* p. 202.
12. *Gazetteer of Bombay City and Island* (1909), Vol. III, p. 66.
13. Forbes, J. (1834): *Oriental Memoirs,* Vol. 1, London, pp. 94-96.
14. Edwards, S.M. (1902): *ibid.,* p. 229
15. *Gazetteer of Bombay City and Island* (1909), Vol. II, pp. 144-145.

16. Postan, M. (1839): *Western India in* 1838, Vol. 1, London, pp. 21-25.
17. Postan, M. (1839): *ibid.*, p. 73.
18. Postan, M. (1839): *ibid.*, pp. 73-87.
19. Postan, M. (1839): *ibid.*, pp. 77-78.
20. Falkland (Mrs.) (1930): Chow-Chow, edited with Introduction and Notes by Rowbinson, H.G., London, Vol. 1, pp. 4-7.
21. Elwood (Mrs.) (1830): "Narrative of an Overland Journey to India", quoted from *Gazetteer of Bombay, City and Island,* Vol. 1, p. 414.
22. Edwards, S.M. (1902): *ibid.*, p. 269.
23. Share Mania: "The trade boom which lasted from 1861 to 1865 was the direct result of the American Civil War which cut off the supply of American cotton to the British textile industry", cited in Edwards, S.M. (1902): *ibid.*, p. 273.
24. *Gazetteer of Bombay City and Island,* (1986), Vol. II, p. 182.
25. Arunachalam, B. (1978): "Bombay, An Exploding Metropolis", Million Cities of India, New Delhi, pp. 76-77.
26. Postan, M. (1839): *ibid.*, p. 73.
27. *Gazetteer of Bombay City and Island* (1909), Vol. II, Bombay, p. 430.

4

POPULATION AND COMMERCE IN BOMBAY WITH SPECIAL REFERENCE TO RETAILING

4.1 Salient Features of the Demographic Scene of Greater Bombay

4.1.1 *Population Growth during the Colonial Period*

When the British King took actual physical possession of Bombay in 1665, its population was estimated around 10,000.[1] Attracted by the inherent potentiality of Bombay, merchants as well as workers flowed steadily into the city in the following decade. Within only eight years (1669 to 1677) of the rule of Sir George Oxenden, the first British Governor of Bombay, its population increased to 50,000. In about a century's time, i.e., by 1780, Bombay's population rose to 113,700 and it further rose to 5.6 lakhs by 1846. In the next hundred years, Bombay's population increased by about 5.7 times and Bombay's area also increased to 62.65 sq kms. The density of population increased to over 10,000 per sq km. It has been reported that by the first ever census of Bombay. Bombay's population reached 7.7 lakhs in 1881, which means a growth of about 15 times in a period of 216 years since the British occupation, or 6.7 per cent average growth per year.[2]

4.1.2 *Bombay's Population in the 20th Century*

Bombay's population has now grown to 9.1 million in 1991, a growth of about 12 times in the period 1891-1991.The continued high growth of Bombay's population is a product of its accretion of urban functions and commerce snowballing with a cumulative effect.

Table 4.1 gives the trends in the population growth of Greater

Table 4.1: Growth of Bomay's Population

Year	Population (lakhs)	Percentage of Decadal Growth Rate	Bombay's Share in urban Population of	
			Maharashtra	India
1901	9.28		25.3	3.2
1911	11.49	23.8	31.3	4.0
1921	13.80	20.1	32.3	4.5
1931	13.98	1.3	28.4	3.8
1941	18.01	28.9	28.9	3.8
1951	29.94	66.2	32.2	4.8
1961	41.52	38.7	37.2	5.4
1971	59.72	43.8	38.0	5.6
1981	82.43	38.0	37.3	5.3
1991	99.10	20.2	32.5	4.6

Source: Census of India (1971), (1981) and (1991)—Maharashtra, General Population Table Series II, Part IIA.

Bombay during the period 1901 to 1991, rate of decadal growth and its relative share in the urban population of Maharashtra and India.

The relative share of Bombay in the urban population of Maharashtra and India has been on a growing and upward trend, over the decades though somewhat stemmed in the latest census decade. However, the population growth rates have been oscillating, reaching a peak in the decade 1941-51 and coinciding with an unanticipated refugee influx into the city, following Independence. After 1951, the rates of growth have been slowing down, but in terms of absolute numbers it has been shooting a rising trend. This growth brought about not only a phsycial growth in the urban area, but also an increasing intensity both in horizontal and vertical expansion of land uses. This abnormally steep rise in number and rate of growth of population cannot be totally explained by natural increase alone, especially in view of the fact that the urban population shows an abnormally high concentration in working age groups and that too essentially of the male population. Migration into the city has played a significant role in its growth in numbers.

From the available data it is revealed that till 1941, migration alone contributed to the growth of Greater Bombay's population. During the decade 1941-1951, out of the total addition to Greater Bombay's population of about 1.2 million (mn.), 950 thousands, or 79.6 per cent, was contributed by migration as against 243 thousands, or 20.4 per cent by natural growth, which is primarily due to refugee influx. Since then, the share of migration in the city's population appears to be slowing down while the percentage share of natural increase for the first time exceeded migration in 1971-81. Out of the total increase of 2.27 mn. during this period, natural growth contributed 1.23 mn. and migration 1.04 mn.

4.1.3 *Migrant Workforce and its Shifts in Status over Decades*

However, in terms of absolute increase, migrant influx is still increasing in number, though natural increase has exceeded the migrant number in the latest decade. Bombay's attraction to the migrants still remains unabated though the increasing pressures on the basic living facilities and amenities and decreasing employment opportunities have started acting as disincentives to further migration to Bombay while possibly other avenues in other urban centres have also started diffusing the migrant streams. This migrant stream especially of work force has contributed much to the growing prosperity and affluence of Bombay. One of the earliest groups of migrants were the trading communities, primarily Gujaratis and Rajasthanis who were encouraged by colonial interests to settle in the city, near the administrative core either within the fortified town or just outside it for the obvious reason of availing of all trade facilities and developed infrastructure.

Over the years with the job opportunity in industrial sector declining, the migrants not only from Maharashtra but also from Bombay's wide catchment area, which included Gujarat, Andhra Pradesh, Karnataka, Tamil Nadu, Kerala, U.P. and Bihar, entered into the informal sector of trade and business. This is reflected in the mushroom growth of pavement shops and in the growth of informal sector, which contributes a substantial share, yet undetermined, to the total volume and value of retail trade in the city.

4.1.4 *Rapid Expansion of Suburbs*

The growth of population has been more rapid in suburbs (eastern as well as western) and even more so in extended suburbs, i.e., Ward P/North and R in the western suburbs and ward T in eastern suburbs. During the last four decades the variations in the comparative growth rates have been more marked. Compared to 4.39 per cent rate of growth in population in the island city during 1941-1951, the suburbs had a very high growth rate (8.52%) of population, and the extended suburbs had 3.75 per cent growth But in 1961-71, the growth rate of population in the city declined to 1.02 per cent while the suburbs had 7.67 per cent and extended suburbs 7.83 per cent rate of growth.[3]

4.2 Temporal Changes in Urban Work Force in Bombay based on Secondary Data Analysis

4.2.1 *Changing Industrial Scene of the 20th Century*

Though early growth of Bombay was built on its overseas commerce with Britain and China, since 1855, cotton mill industry growing fast, took a larger and larger share of its work force. A further diversification of industries since the 1920s saw a rising industrial eminence of the city. The development of the work force of Bombay has, however, been more stable and steady since the, 1940s compared to the previous decades of the 20th century. The favourable economic policies of the Government that helped to boost industrialisation, the location of the headquarters of several national financial institutions and major banks in Bombay, the advantage of a colonial port city with a long heritage of trade, commerce and entrepreneurial zeal of its people—all helped the faster growth of Bombay as the commercial capital of India. The fast expansion as well as diversification of the industrial sector became pronounced after World War II. The textile industries which accounted for about 73 per cent of Bombay's factory employment in the 1920s came down to 36 per cent in 1971. The metal and engineering sectors have been expanding more rapidly than the textile sector. Other industries like paper and printing, chemicals and drugs, food products, tobacco, leather, furniture and timber, jewellery and ceramic products and a host of other products have been added to Bombay's growing industrial sector.

The number of working factories increased by 7.4 times from 1,078 in 1941 to 7,971 in 1989 while the number of workers engaged in the factory industry and other small-scale industry increased from 314 thousands in 1941 to 547 thousands in 1961 and further to 802 thousands in 1981.

4.2.2 *Growing Importance of Commerce Accompanying Industrial Diversification*

Alongside the growth of the secondary (industrial) sector, the tertiary service sector also has been gaining prominence in Bombay's economy with increasing number of work force engaged in it. Thus, the employment pattern in Bombay has been gradually showing a sign of structural shift from secondary sector to tertiary sector.

Table 4.2 indicates this trend. It is seen from the table that the share of the industrial sector in the city's work force has increased from

Table 4.2: A Comparison of Number of Workers per Thousand in Industrial and Service Sector in the City of Bombay (1901-1981)

Year	Total Workers	Industry	% of industrial workers to total workers	Trade and commerce	% of workers in trade to total workers	Workers in transport	% of workers in transport to total workers
1901	1000	332	33.2	133	13.3	89	8.9
1911	1000	309	30.9	135	13.5	129	12.9
1921	1000	293	29.3	146	14.6	85	8.5
1931	1000	285	28.5	141	14.1	84	8.4
1951	1000	372	37.2	210	21.0	91	9.1
1961	1000	408	40.8	180	18.0	112	11.2
1971	1000	422	42.2	222	22.2	107	10.7
1981	1000	414	41.4	218	21.8	100	10.0

Source: Census of India 1961, Vol. X Maharashtra Part IC, Subsidiary Tables.
Census of India 1971 and 1981 Maharashtra Part IIA & B (1), General Economic Tables B1-B6.

33.2 to 41.4 per cent during the period 1901 to 1981, while during the same period the share of commerce (including hotel and cinema workers) has increased from 13.3 to 21.8 per cent and that of transport marginally from 8.9 to 10 per cent. This clearly demonstrates that with the growing importance of industrial sector, the commercial importance of the city (which started its foundation of prosperity entirely on its commerce) has not declined but on the other hand has been reinforced and augmented, to grow at a faster rate as reflected by the trends in work force growth.

The analysis could not be carried out for earlier decades for want of comparable data.

4.2.3 *Spatial Analysis of Urban Work Force Distribution Patterns*

In a study of the occupational structure of the resident population of different sections of Bombay based upon 1971 Census, Sahasrabudhe[4] observed that manufacturing constitutes the first ranking occupation in most of the sections of Greater Bombay, although its proportion increases from outer suburbs towards the city core. All sections in Wards E, F and G in the centre and north of the island city and practically the whole of the suburbs except the sections along the western shore and Aarey have manufacturing as first ranking occupation. The old textile core in the centre of the city has 50 per cent to 67 per cent of the population engaged in manufacturing.

Trade and commerce rank first in 12 sections lying exclusively in B and C wards and eastern part of D ward. The area coincides with the areas whose land-use characteristics concentrate wholesaling, semi-wholesaling and the older retail clusters.

It may also be noted that a fairly consistent high degree of association is observed between manufacturing as first ranking and trade and commerce as second ranking occupations over most of Greater Bombay. Trade and Commerce is ranking third only in eastern suburbs and some parts of the inner western suburbs.

4.3 Demographic Characteristics of the Retailing Sector in Greater Bombay

4.3.1 *The Nature of Analysis*

Having examined the broad general aspect of work force in Greater Bombay it is now important to analyse some relevant aspects of employment in retail trade within the trade and commerce sector of Greater Bombay. Accordingly, in the following section, an analysis is made of the following:

(i) Growth of population and retail employment during the period 1971 ato 1983.
(ii) Retail Employment Participation Ratio (REPR);
(iii) Density of Population Engaged in Retailing (DPER);
(iv) Index of Retail Employment (IRE); and
(v) Intensity of Retail Employment (INRE)

4.3.2 *Data Base*

Data on employment in retail establishments for the years 1971 to 1983 were collected and compiled from various Ward Offices of Bombay Municipal Corporation (BMC) and from BMC's Central Office for shops and establishments. The BMC keeps a record of number of employees for the issuance as well as renewal of licences for shops. Since the break-up of retail employment is not available in the published census data, a recourse to the collection of the retail employment data for the licensed shops from the unpublished office data of the BMC had to be employed. Since this information is maintained by BMC for employees' welfare measures like mandatory one day holiday in a week and security measures, it is likely that the data is an underestimate of actual retail employment. Further these datas totally exclude retailing employment in the informal sector which, as indicated earlier, explains about a third of the total retailing employment. The general pattern of distribution of retail trade and employment in the urban complex becomes clear from the analysis of the mapped data of the 1971 and 1983 municipal records.

4.3.3 *Association between Growth of Population and Retail Employment during the Period 1971 to 1983*

According to retail employment data collected from BMC, employment in retail sector in Greater Bombay as a whole increased from 115.36 thousands in 1971 to 253.77 thousands in 1983, indicating an average annual growth of about 10 per cent.

In the island city (comprising of BMC's administrative wards, A, B, C, D, E, F, and G), retail employment increased from 97.30 thousands in 1971 to 198.13 thousands in 1983, indicating an average annual growth of 8.64 per cent.

Similarly, in the Western suburbs (comprising of wards H, K, P,R) retail employment increased from 12.02 thousands in 1971 to 36.20 thousands in 1983, indicating an average annual growth rate of 27.93 per cent. In the eastern suburbs (comprising of wards, L, M, N, S and T), retail employment increased from 6.04 thousands in 1971 to 19.46 thousands in 1983, indicating an average annual growth rate of 18.51 per cent. Table 4.3 shows these trends.

It is clear from Table 4.3 that during the period 1971 to 1983, the suburbs of Bombay and more so the Western suburbs experienced a far higher growth in retail employment than the island city. During this period, which more or less corresponds with the 1971 and 1981 population censuses, the rate of growth of population was also considerably higher in the suburbs (67.62 per cent for the western suburbs and 75.79 per cent for the eastern suburbs) *vis-a-vis* a relatively small 7 per cent growth in the island city. For the whole of Greater Bombay, population growth was a little over 38.08 per cent during 1971 to 1981 corresponding to 120 per cent growth of retail employment during 1971 to 1983.

The comparatively modest rate of growth of population during 1971-1981 in the island city is also related with the significant phenomenon of declining city core as note earlier. With the expansion of the city within and around the Fort to the island city and then progressively to the western and eastern suburbs, the retail trade and retail employment also followed suit.

Over time, the horizontal growth (i.e., in terms of spatial expansion) of commercial activities gradually spread to the expanding suburbs, first

TABLE 4.3: Growth of Population and Retail Employment in Greater Bombay

		Population (in '000s)		Percentage Growth of Population in 1981 over	Retail Employment (in '000s)		Percentage Growth of Retail Employment in 1983 over	Average Annual Growth of Retail Employment during	Retail Employment Participation Ratio (REPR)	
		1971	1981	1971	1971	1983	1971	1971-1983	1971	1983
1.	Island City (Ward: A, B, C, D, E, FS, FN GN, GS)	3,070.38	3,285.05	7.00	97.30	198.13	103.63	8.64	0.0317	0.0603
2.	Western Suburbs (Wards :HW, HE, KE, KW, PN, PS)	1,705.49	2,858.77	67.62	12.02	36.20	201.16	16.76	0.0070	0.0127
3.	Eastern Suburbs (Wards: L,M, N, T & S)	1,194.71	2,100.23	75.79	6.04	19.46	222.18	18.51	0.0051	0.0093
4.	Greater Bombay (1+2+3)	5,970.58	8,244.05	38.08	115.36	253.79	120.00	10.00	0.0193	0.0308

Note : Detailed retail employment data for all wards were not available for 1981 and 1982. 1983's, data collected from the B.M.C., were used for computing the REPRs over the population figures of 1981. The retail employment in 1981 will be less than that of 1983, changes in REPRs will be insignificant.

Source: Retail employment data were collected from B.M.C.

for residential purposes and then for industrial as well as trade functions though in a more sporadic manner. But the city core, being subject to intense landuse pressure and competition, experienced limited vertical growth of commercial activities as well, and thus led to its increased intensity.

It is only in the recent past that the suburbs have been experiencing relatively higher rate of growth in population as well as higher growth in retail employment. Yet, it cannot be denied that the geographical inertia persists in favour of the older retailing areas of B and C wards, notwithstanding the fact that recent residential growth is primarily in the suburbs, clearly indicating that the development and growth of retailing in the suburbs lags behind and is slow to follow the haphazard and unplanned sprawl of residential consumer areas. Shopping areas are not planned simultaneously with residential growth inputs and this has led in many cases to dispersed and scattered distribution of shopping for the basic and ubiquitous daily needs of retailer service.

4.3.4 *Retail Employment Participation Ratio (REPR)*

It is interesting to observe from Table 4.3 that retail employment participation ratios (REPR) have almost doubled from 0.0317 in 1971 to 0.0603 in 1983 in the island city, signifying higher retail employment and, therefore, much higher retail functions in the island city. In the suburbs, these ratios are low, varying only in the 3rd and 4th decimal point positions of the ratio. The REPR, although very low in magnitude for the western suburbs, increased from 0.007 in 1971 to 0.0127 in 1983, or almost doubled. As for the eastern suburbs, REPR grew from 0.0051 to 0.0093, again almost doubled during this period. As for the whole city, it grew from 0.0193 to 0.0308 during 1971-1983, or almost doubled. The apparent imbalance in retail employment distribution in island city *vis-a-vis* the suburbs as indicated by the highly skewed distribution of REPRs is related to the city's nucleus, the city core wherein the commercial activities were and still are concentrated. At the same time, the relatively high growth rate of retail employment in the suburbs, albeit with low (till now) REPRs, does indicate the future propect of higher retail function, including, of course, higher retail employment. This is further corroborated by the fact that the city core is declining and the growth rate of population being very low in the island city, the suburbs indeed will be the future growth areas for commercial activities including, naturally, retail traing and, therefore, retail employment.

TABLE 4.4: Ward-Wise Distribution of Population and Retail Employment in Greater Bombay

Wards	Area in Km^2	Population (in '000s) 1981	Growth of Population during 1971-81	Density of Population ('000s per km^2)	Total Workers in 1981 (in '000s)	Percentage share of Workers in Ward to Total Workers	Population Engaged in Retailing 1982 (in '000s)	Density of population Engaged in Retailing (DPER) (in '000s per km^2)	Index of Retailing Employment (IRE) (in %)	Percentage of Retailer in Relation to Total Retail Employment	Intensity of Retail Employment (InRE) L.Q. Values
(1)	(2)	(3)	(4)	(5)	(6)	(7)	(8)	(9)	(10)	(11)	(12)
A	10.46	168.22	– 8.6	16.68	71.30	2.50	28.62	2.73	4.84	13.00	1.46
B	2.46	147.31	– 15.9	59.88	53.53	1.88	16.52	6.71	11.92	6.51	0.73
C	1.78	270.71	– 13.4	152.08	112.34	3.94	55.77	31.33	55.65	21.97	2.50
D	7.21	445.40	– 0.7	61.77	149.38	5.21	55.31	7.67	13.62	21.79	2.44
E	6.79	455.71	– 1.5	67.11	193.81	6.80	11.35	1.67	2.96	4.47	0.50
F(S)	8.55	485.72	15.0	56.80	169.95	5.96	7.03	0.82	1.45	2.77	0.31
F(N)	12.60	328.98	37.0	26.10	112.26	3.94	5.30	0.43	0.76	2.08	0.23
G(N)	7.37	445.96	28.5	60.51	153.04	5.37	10.23	1.38	2.45	4.03	0.45
G(S)	10.48	537.04	12.4	51.24	205.88	7.22	8.00	0.76	1.35	3.15	0.35
Total Island City	67.70	3,285.05	7.0	48.52	1,221.49	42.85	198.13	2.87			

(Contd.)

TABLE 4.4 *(Contd.)*

(1)	(2)	(3)	(4)	(5)	(6)	(7)	(8)	(9)	(10)	(11)	(12)
H(W)	6.58	20.8	20.8	35.16	101.61	3.56	6.13	0.93	1.65	2.41	0.27
H(E)	12.98	43.2	43.2	36.62	126.58	4.44	2.22	0.17	0.30	0.87	0.09
K(E)	23.59	54.7	54.7	22.43	180.68	6.34	9.48	0.40	0.70	3.73	0.42
K(W)	22.14	72.6	72.6	17.64	137.47	4.82	5.57	0.25	0.44	2.19	0.24
P(N)	43.98	78.9	78.9	8.48	119.48	4.12	3.70	0.08	0.14	1.45	0.16
P(S)	21.22	74.7	74.7	13.95	103.89	3.64	2.73	0.12	0.21	1.07	0.12
R	77.56	38.3	38.3	7.24	198.91	6.98	6.37	0.08	0.14	2.50	0.28
Total Western Suburb	208.05	2,858.77	67.6	13.74	965.62	33.88	36.20	0.16			
L	13.46	63.0	63.0	32.24	145.55	5.11	5.63	0.37	0.66	1.98	0.22
M	54.92	78.8	78.8	10.30	169.92	5.96	4.29	0.18	0.14	1.69	0.18
N	39.47	72.5	72.5	15.18	274.44	9.63	5.12	0.17	0.30	2.01	0.22
S	15.97	110.7	110.7	17.43	N.A	N.A	2.71	N.A	N.A	N.A	N.A
T	34.45	77.8	77.8	6.46	73.33	2.57	2.31	0.14	0.24	1.97	0.22
Total Eastern Suburb	158.27	2,100.23	75.8	16.27	663.24	23.27	19.46	0.12			

Source : 1. Primary Census Abstract, Greater Bombay, 1981.
2. Retail Employment Data collected from B.M.C.'s unpublished record (1982).

Note: 1. Ward 'S' has been recently formed: hence, Retail Employment Data are not available.
2. "S", "N", "W" and "E" within parentheses after the names of wards indicate south, North and West and East respectively.

4.3.5 *Density of Population Engaged in Retailing (DPER)*

That there is high concentration of retail trade and retail employment becomes clear from a study of density of population engaged in retailing (DPER). For this purpose, *inter alia,* a compact table (Table 4.4) has been constructed which also shows that density of retail employment is very high in ward C (comprising of section Khara Talaco, Kumbharwada, Bhuleshwar, Market, Dhobi Talao and Fanaswadi) followed by D (Khetwadi, Tardeo, Girgaum, Chaupathy, Walkeshwar and Mahalaxmi, B (Mandvi, Chakala, Umarkhadi and Dongri), A (Upper, Middle and Lower Colaba, Fort South and North and Esplanade), E (Mazgaon, Tadwai, First and Second Nagpada, Kamthipura and Byculla), G (North (Dadar and Mahim), H West (in western suburb-Bandra, Khar, Santacruz etc.), F South (Parel, Sewri and Naigaon) and G South (Prabhadevi, Worli, Chinchpokli, Love Grove etc.) in that order.

The other wards have a very lower level of concentration of retailing.

All these wards with high density of retail employment are located in the island city (except H west ward which is in the inner western suburbs, close to the island city across its Mahim Causeway).

The DPER is highest—31.33 thousand per square kilometre (km^2)—in the case of C ward, followed by D ward with 8.34 thousand per km^2, B ward with 6.71 thousand per km^2 and A ward with 2.52 thousand per km^2. There is a strikingly wide gap in DPER values between ward C, where it is highest, and ward D, where it is second highest. This singles out ward C for the most dense concentration of retail employment and a distinctive primacy. With an area of only 1.78 km^2 it is smallest in size among all wards with the highest retail employment.

It is also seen from Table 4.4 that in the suburbs, DPERs are considerably lower than those in the island city, or the city core, falling in certain specific sections of wards A, B, C and D. This also supports the observation made earlier (in para 4.3.3): namely, the significantly higher population and retail employment growth rate in the suburbs as against modest growth of these two parameters in the island city. The relatively low DPERs in both the western and eastern suburbs implies that there is future prospect of higher retail trade activities accompanied by higher retail employment. This also therefore implies that in future DPERs are likely to increase in the western and eastern suburbs alongwith the continuation of higher population growth.

Table 4.5: Wards Showing Higher Concentration of Retail Employment

Wards	DPER ('000 employment per km²)	IRE (%)	InRE (L.Q. Value)
C	31.33	55.41	2.50
D	8.34	14.75	2.44
B	6.71	16.86	0.73
A	2.51	4.43	1.46
E	1.53	2.71	0.50
G. North	1.32	2.33	0.45
H. West	0.93	1.64	0.27
F. South	0.82	1.45	0.31
G. South	0.76	1.34	0.35
F. North	0.43	0.76	0.23
K. East	0.40	0.71	0.42
K. West	0.25	0.44	0.24
R	0.08	0.14	0.28

Notes:

1. Table 4.5 has been constructed for highlighting the wards with Retail Employment Concentration.
2. Wards have been arranged in the descending order of the value of:
   (i) Density of Population Engaged in Retailing (DPER)
   (ii) Index of Retail Employment (IRE)
   (iii) The Intensity of Retail Employment (InRE) based on Location Quotient (LQ) Values does not always strictly follow the above order.
3. Wards C, D, B and A have distinctly higher concentration of retail employment and consequently retail trade. The rest of the wards have relatively lower values of DPER, IRE and InRE.

4.3.6 *Index of Retail Employment (IRE)*

The index of retail employment (IRE) for a particular ward shows the relationship of retail employment with the total DPERs of various wards of Greater Bombay.

A close look at Table 4.5 will show that the order of wards with decreasing figures of DPER is highly consistent with the order of values of IREs. For example, ward C has the highest value of DPER (31.33 thousands of poeple per km² engaged in retail trade) as well as the highest value of IRE (55.41%) followed by D ward with the record highest DPER (8.34 thousand per km²) and the second highest value of IRE (14.75%) followed by B ward with the third highest value of DPER (6.71 thousand

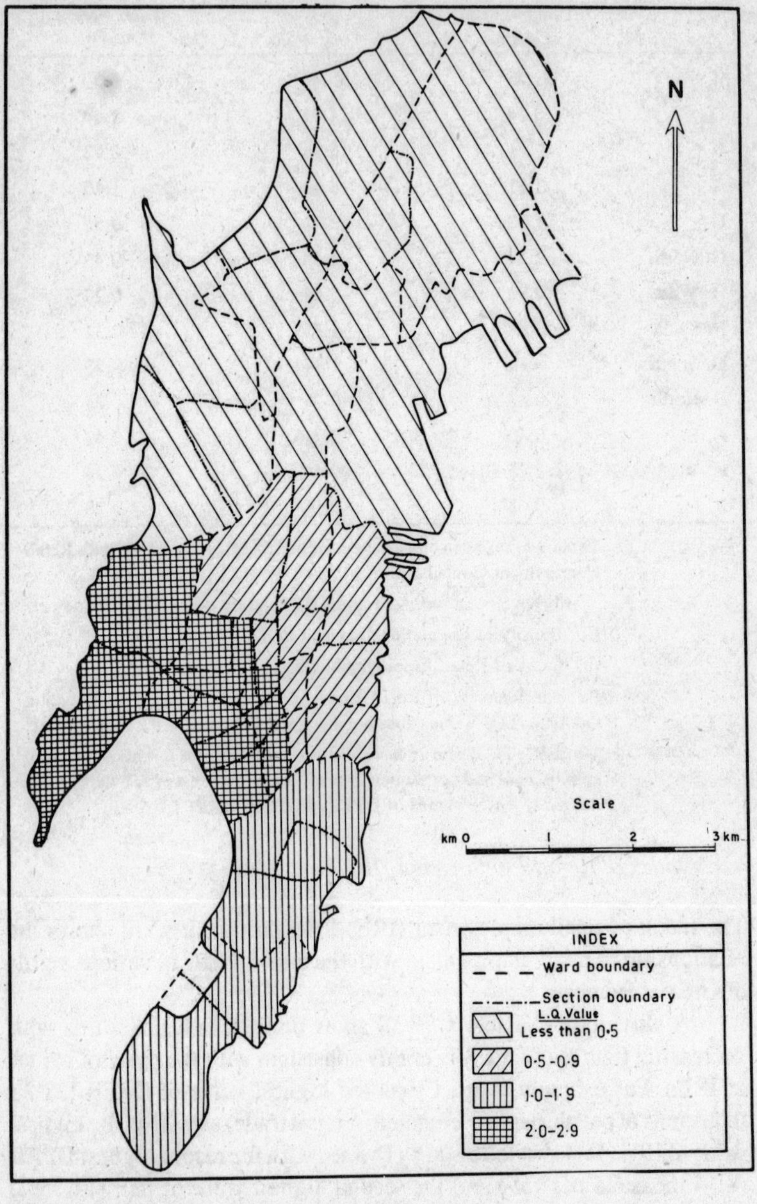

MAP 4.1: Intensity of Retail Employment (Bombay City)

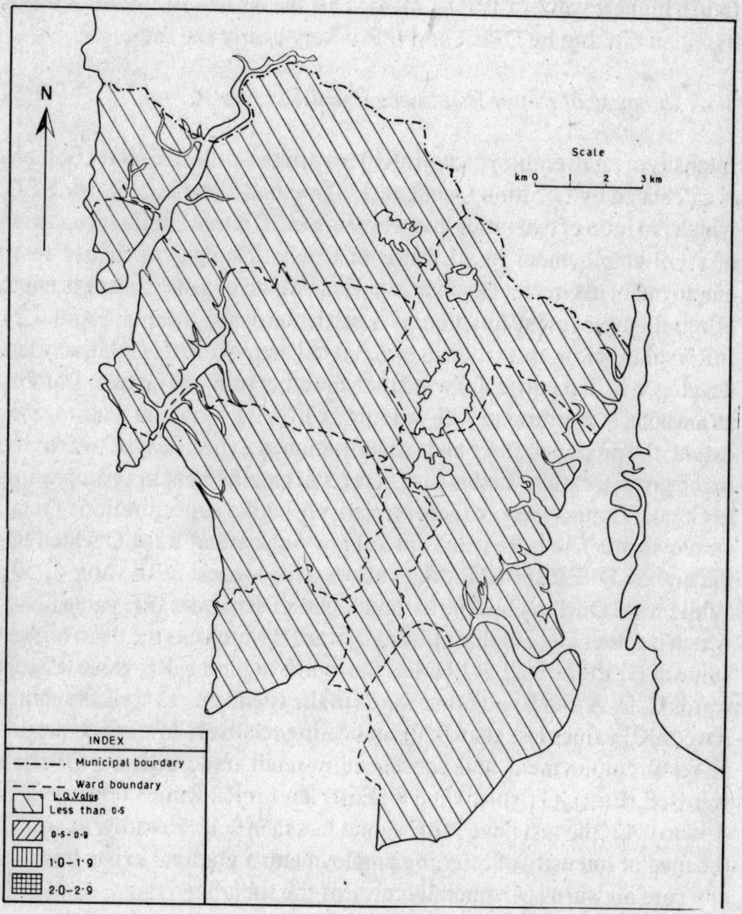

MAP 4.2: Intensity of Retail Employment (Suburbs of Bombay)

per km^2) and the third highest value of IRE (11.86%) followed by A ward with the fourth highest DPER value (2.51 thousand per km^2) and the fourth highest value of IRE (4.43%). This means that the order of wards based on ranking by DPER and IPR is very nearly the same.

4.3.7 *Intensity of Retail Employment (InRE)*

Intensity of retail employment (InRE) in various wards of Greater Bombay is calculated by Location Quotient (LQ) method.[5] The idea of using LQ, which is a ratio of two ratios, in measuring InRE is that similar proportion of retail employment in all wards of Greater Bombay to that of retail employment in Greater Bombay as a whole would be a measure of strength of retail employment function of a particular ward (Map 4.1 and 4.2). InRE value below unity indicates that retail employment is relatively less developed in that particular ward than the other wards of Greater Bombay as a whole. According to R.B. Andrews,[6] LQ also shows, at least to some extent, the preponderance and specialisation of a specific type of activity, in the given location. In this case, it is retail employment in certain wards of Greater Bombay where there is relatively higher concentration of retail employment. The InRe values in Table 4.5 show that ward C which has the highest DREP and IRE values also has the highest InRE value (2.50). While ward D which has the second highest DREP and IRE value (2.44) which is rather close to that of ward C, ward B which has the third highest value of DPER as well as IRE has the fourth highest InRE value. Except wards C, D, A and B, all other wards in the island city as well as suburbs have InRE values less than 0.50, indicating relatively low concentration of retail employment and consequently retail trade activity. Of these wards, E (East), G (North) and K (East) have InRE value in the range of 0.50 to 0.42. the rest have InRE values less than 0.42. Broadly, as seen in the map of intensity of retailing employment, a gradient exists from the city core outwards peripherally towards the suburbs.

A comparison of the vlaues of DPER and IRE and InRE in the different wards shows clearly that the wards with high concentration of retail employment have high concentration of retail trade.

It can now be summarised:

(i) At the ward level, the highest concentration of retail employment is in ward C, where density index and intensity of retail employment is maximum. In ward C, Bhuleshwar and Market (sections 12 and 13) have got overwhelming importance of this trade sector in employment which is associated with transport and storage where the service is the third ranking occupation;

(ii) This retail peak area is adjoined by B and D wards, but with a sharp fall of density, index and intensity from C as apparent from Table 4.5. This intra-zonal variation of retail employment within the Bazaar zone can be explained by the fact that in B ward, specially in Mandvi-Chakala, wholesaling is predominantly associated with warehousing and transportation. In ward D, retailing is intimately associated with service-seeking old residential clusters;

(iii) North and South Colaba in ward A show trade and commerce as the second ranking occupation associated with administration, but in Fort North (section 4), trade and commerce is the first ranking occupation; here the percentage of retailers to total retailers in the city population is higher (13%) than in the wholesaling district of B ward (6.51%);

(iv) Retailing density and intensity reach the lowest values in the industrial labour core of Parel, Chinchpokli and Naigaum; in these area, trade and commerce has only the third ranking position;

(v) Retailing employment has a subsidiary low peak in Dadar that has an intra-urban transport nodal location at the convergence of the rail commuting of all the rail corridors with considrable transhipment between them; and

(vi) In the suburbs, retailing as an employment sector is still in this infancy of development. Hence, retail index is less than one in all the wards except H(W)-Bandra and Santacruz (West).

REFERENCES

1. The figure 10,000 is Fryer's estimate quoted from *Gazetteer of Bombay City and Island* (1909) Vol. 1 p. 15.
2. *Gazetteer of Bombay City and Island* (1909); Vol. 1 pp. 156-158.
3. Revati, K. (1977): Pattern of Population Distribution in Bombay, Unpublished M.A. Dissertation, Department of Geography, University of Bombay
4. Sahasrabudhe, S.S. (1975): Occupational Structure of Residential Area of Greater Bombay—A Spatial Review, Unpublished M.A. Dissertation, Department of Geography, University of Bombay.
5. Location quotient represents a popular method for determining the ratio, sometimes called the "approximation technique". It grew out of the work of Home Hoyt:
Hoyt, H. (1961): The Utility of the Economic Base Methods in Calculating Urban Growth,

Land Economic, Vol. 37, pp. 51-58.

6. Andrews, R.B. (1953): Mechanics of the Urban Economic Base, *Land Economics,* Vol. 29, pp. 343-350.

   Reprinted in Ralph, W.P. (ed.): The Techniques of Urban Economic Analysis (1960. Chandler-Davis, West Trenton, New Jersey.

5

MORPHOLOGY AND STRUCTURE OF RETAILING IN BOMBAY

5.1 The Functional Amalgam of Greater Bombay

Earlier researchers like Nissel[1], who have studied Bombay, have identified, based upon empirical field investigation, the broad zones of functional morphology in the island city. Adopting the same and extending the concept to the suburban zone as well, the main commercial functional areas are identified, based on empirical observations. The boundary limits are adjusted to section boundaries, as secondary data compilation from administrative records is possible only in terms of such sections. Accordingly, a map is constructed to show the concentration of different functions in the city (Map 5.1)

(i) Sections 3, 4 and 5 consisting of Fort North and South, and Esplanade coinciding with the original European town and the more recent Backbay Reclamations at the Nariman Point is the city nucleus; or city centre that can be termed the CBD. It is the administrative - cum - business core, the main banking and financial district and the home of publications and printing. It has a strongly active retail core in both formal and informal sectors along the main roads converging on the two rail terminals at V.T. and Churchgate and the bus terminals of Hutatma Chowk and Mantralaya (State Secretariat).

(ii) To the north of the Fort, adjoining the docks and the rail yards, is the commercial core, with wholesaling predominant in sections 6 and 7, that is, Chakala, Mandvi and Masjid, gradually grading into semi-wholesaling and retailing westward in sections 8, 10, 12 and 13 that is Bhuleshwar,

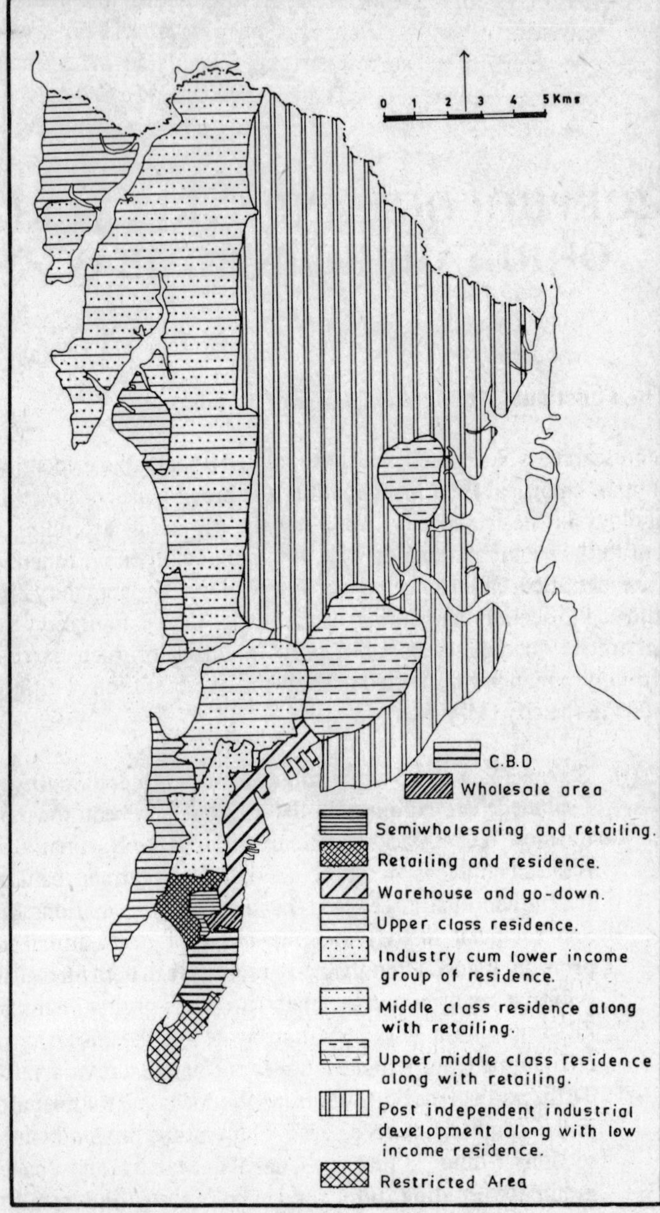

MAP 5.1: Functional Amalgam of Greater Bombay

Khara Talao, Kalbadevi and Umarkhadi. In these sections, there is a strong association of wholesaling and retailing activities, transportation and warehousing. The area is characterised by narrow streets, dilapidated old housing, heavy pedestrian traffic and slow-moving hand cart movement, competing with vehicular truck traffic and a high intensity of business activity round the clock. Westward, retailing has permeated and penetrated along the main roads in ribbons into the old residential cores of Girgaum and Chowpatty sections;

(iii) To the south of the Fort and its CBD complexes is a mixture of low-income residential zone associated with dock workers, the retail ribbon along the Colaba-Causeway, and the upper middle and upper class exclusive residential areas like Parsi colonies (Cusrow Baug) and the newer skyscraper residential complex of Cuffe Parade;

(iv) Central Bombay is dominated by big industry, low income industrial workers' residential quarters freely intermingle with the mill areas. Retail shopping is along the main road fronts;

(v) In the north of the city island, in Dadar, Sion, Matunga and Mahim is a middle class residential area. Here retailing converges on the rail-heads and the main road chowks.

(vi) So too, in the western and eastern suburbs, rail-head, commuter-convergent retail shopping characterises these areas. The link roads between the central and western suburbs also have the retailing function.[2]

5.2 A Morpho-Structural Study of Retailing

5.2.1 *Data Base for Study of Floor-space Intensities*

A morpho-structural study based on floor-space-wise intensities of certain functions like (i) wholesaling, (ii) retailing (iii) residential function was carried out to show the nature and intensity of concentration of all these functions in Greater Bombay.

For this analysis, section-wise secondary data have been obtained from the records of Bombay Municipal Corporation (BMC). The BMC has conducted its latest survey (1980) of floor space occupied by various functions like wholesaling, retailing and residential functions. Floor space indices (FSI) assigned to each section/ward by the BMC have been used for comparison.

The data on floor-space occupation by all the different functions in various sections of the BMC, for the island city and the various wards of the suburbs have been compiled from unpublished records of the BMC. The floor - space occupation data of functions for the island city have been collected and recorded for each of the sections of the city ward of the BMC However, for the suburbs, the BMC survey has taken into consideration and collected data from areas upto a distance of 1.6 km (one mile) on either side of the railway line and not in terms of sections as in the case of island city.

5.2.2 *Index of Floor - Space Intensity*

Based on the BMC data, floor-space-wise intensities have been worked out using Location Quotient (LQ) method. For example, the index of wholesale intensity in the island city (based on floor-space occupation) is calculated thus:

$$\frac{\text{Percentage of floor space occupied by the retail trade in the city section}}{\text{Percentage of floor space occupied by the wholesale trade in the whole island city}}$$

Similarly, the wholesale/retail intensity for the suburbs has been calculated thus:

$$\frac{\text{Percentage of floor space occupied by wholesale/ retail trade in the suburban ward}}{\text{Percentage of floor space occupied by wholesale/ retail in the suburbs as a whole}}$$

The principle underlying the use of LQ which is a ratio of two ratios would give a measure of relative locational strength or concentration of a specific function like wholesale trade, retail trade and residential function in a particular section/ward *vis-a-vis* the city/suburban zone as the case may be. L.Q. values below unity would indicate that the function is relatively less developed or has a poorer concentration in a particular section or ward than in the entire city/suburbs as a whole. Similarly, L.Q. values more than unity would indicate a relatively more pronounced concentration and these in comparison with the F.S.I. values obtaining in

TABLE 5.1: (A) Ward-wise F.S.I. Values of Different Urban Functions
Suburb (Ward-wise)

Ward	Name of the area	Retail Intensity	Wholesale Intensity	Residential Intensity
H	Bandra, Santacruz, Khar, Kalina	0.67	0.00	1.80
K	Vile Parle (E,W) Andheri, Versova, Jogeshwari	0.59	0.00	1.70
P	Goregaon (E,W) Malad Madh, Dindoshi	0.82	0.00	1.60
L	Kurla	0.71	0.00	1.40
R	Borivli, Poisar, Dahisar	0.60	0.00	1.90
M	Chembur, Trombay	0.58	0.00	1.70
N	Ghatkopar, Vikhroli, Powai	0.61	0.00	1.20
T	Mulund	0.56	0.00	1.40

(General F.S.I. Values for the suburban wards are not available.)

the given section would be a further indicator of the relatives strength of a given function in comparison to other functions in the section/ward.

5.2.3 *Spatial Variations of F.S.I. Values in the City*

Section-wise F.S.I. values are compiled in Map 5.2. High F.S.I. Values are recorded in Section 4, Fort North (3.81), Section 7, Chakala (3.74), Section 13, Market (3.46) and Section 10, Khara Talao (3.23). Least F.S.I. values are seen in Sections 29, Sewri (0.31) and 32, Sion (0.30). Higher values of F.S.I. indicate higher land rent resulting in a concomitant vertical growth of built-up areas. The F.S.I. values are the highest in the administrative-cum-banking city centre of Fort South and North, as well as sections of the commercial core of the C ward. In general, the F.S.I. values decline from the city core northwards and with increasing distance from the two main (central - western) rail corridors, the least values being recorded in Sion-Wadala.

The calculated intensities of occupation of various urban functions, section-wise for the city and ward-wise for the suburbs, have been shown in Table 5.1 (A) and 5.1 (B) and the spatial spread of the intensity pattern, plotted on Maps (5.2 to 5.8 on a cholropleth scheme.

TABLE 5.1: (B) F.S.I. Values for Different Urban Functions of the Sections of the City

Sr. No.	Sections	F.S.I. (General)	Retail Intensity	Wholesale Intensity	Residential Intensity
1.	Upper Colaba	1.68	0.53	0.00	0.40
2.	Middle, Lower Colaba	2.33	0.80	0.13	0.80
3.	Fort South	1.98	0.77	0.40	0.50
4.	Fort North	3.81	6.50	3.16	0.50
5.	Esplanade	1.37	0.60	0.60	0.70
6.	Mandvi	1.59	0.69	6.07	0.30
7.	Chakala	3.74	1.98	8.27	0.70
8.	Umerkhadi	2.12	1.48	0.98	1.00
9.	Dongri	0.66	1.44	0.67	0.70
10.	Khara Talao	3.23	3.51	1.63	1.05
11.	Kumbharwada	2.59	1.70	2.80	0.90
12.	Bhuleshwar	2.98	1.75	6.15	0.90
13.	Market	3.46	2.24	11.39	0.50
14.	Dhobi Talao	2.54	1.46	2.47	1.01
15.	Fanaswadi	1.87	0.97	1.46	1.16
16.	Khetwadi	2.22	2.05	1.22	1.06
17.	Tardeo	1.58	1.63	0.85	1.10
18.	Girgaum	1.86	1.23	0.26	1.10
19.	Chowpatty	2.23	1.66	1.67	0.80
20.	Walkeshwar	1.18	0.17	0.00	1.40
21.	Mahalaxmi	1.38	0.41	0.15	1.20
22.	Margaon	0.71	0.69	0.70	0.50
23.	Tadwadi	1.07	0.57	1.07	0.50
24.	1st Nagpada	1.19	0.29	1.25	0.60
25.	2nd Nagpada	1.80	1.87	1.30	1.50
26.	Kamathipura	2.02	3.45	0.00	0.90
27.	Byculla	1.05	0.85	0.17	0.80
28.	Parel	1.01	0.84	0.00	0.90
29.	Sewri	0.31	0.45	0.00	0.60
30.	Naigaon	1.07	0.82	0.30	0.70
31.	Matunga	1.11	0.67	0.00	1.30
32.	Sion	0.30	0.45	0.10	1.20
33.	Dadar	0.53	0.89	0.12	1.10
34.	Mahim	1.02	0.64	0.00	0.90
35.	Prabhadevi	0.86	1.00	0.00	0.70
36.	Worli	1.14	0.42	0.00	0.50
37.	Chinchpokli	0.75	0.71	0.10	0.40
38.	Love Grove	0.43	0.34	0.00	0.80

5.2.4 *Intensity of Wholesale Trade (IWT) Function*

As would be seen from Maps 5.3 and 5.4, intensity of floor-space occupied by wholesale trade is highest (11.39) in section 13—the Market area in the island city, followed by section 7—Chakala (8.27); section 12—Bhuleshwar (6.15); section 6—Mandvi (6.07); section 4—Fort North (3.16): secton 11 - Kumbharwada (2.80); Section 14—Dhobi Talao (2.47); section 19—Chowpatty (1.67); section 10—Khara Talao (1.63); section 15 - Fanaswadi (1.46): section 25—second Nagpada (1.30); section 24—First Nagpada (1.25); section 16—Khetwadi (1.22) and section 23—Tadwadi (1.07). These are the sections where the L.Q. value is more than unity. Of these, Market, Chakala, Bhuleshwar, Mandvi, Fort North, Kumbharwada have L.Q. values higher than even the sectional F.S.I. value, indicating a spatial occupation concentration higher than the local F.S.I. standard. Though 1st Nagpada has got a lower L.Q. value, the value is higher than the F.S.I. value, while in Dhobi Talao, Khara Talao, Chowpatty and 2nd Nagpada, in spite of moderately higher L.Q. value of IWT, it is below the local FSI values. FSI being determined on the basis of land values obtaining, the L.Q. values of IWT become a measure of the locational advantages, or otherwise, for a given function in a specific section. Thus, if in spite of high F.S.I. values, the high L.Q. values of I.W.T. in Market, Chakala, Bhuleshwar, Mandvi, Fort North and Kumharwada are higher than F.S.I. it means a distinct locational/site advantage, which the wholesaling function derives, that makes it stick to the area. In the specific instance, the nearness to the cargo terminal of the port-docks, the Wadi Bunder rail terminal and the lorry terminal on the one hand and the warehousing facility closeby on the other, seem to explain the situation. On the other hand, in case of 1st Nagpada, in spite of a low F.S.I. value, the slightly higher than F.S.I. L.Q. value of I.W.T. is obviously due to specific type of commodity wholesaling that cannot afford higher F.S.I. values and land rents, but would yet like to derive advantage of proximity to it in lesser land value areas - in this case, perishable wholesaling. In case of Dhobi Talao, Chowpatty, Khara Talao and 2nd Nagpada areas, the moderately high L.Q. value though lower than F.S.I. is positively a result of peripheral extension from the core areas outwards as spill over of wholesaling of specific, select commodities. In all other sections, wholesale trade is either negligible or altogether absent. The suburbs have a conspicuous absence of wholesaling. However, in Kurla, a wholesale market for scrap is slowly emerging probably anticipating the shifting of a part of the wholesaling complex to the Bandra-Kurla reclamation.

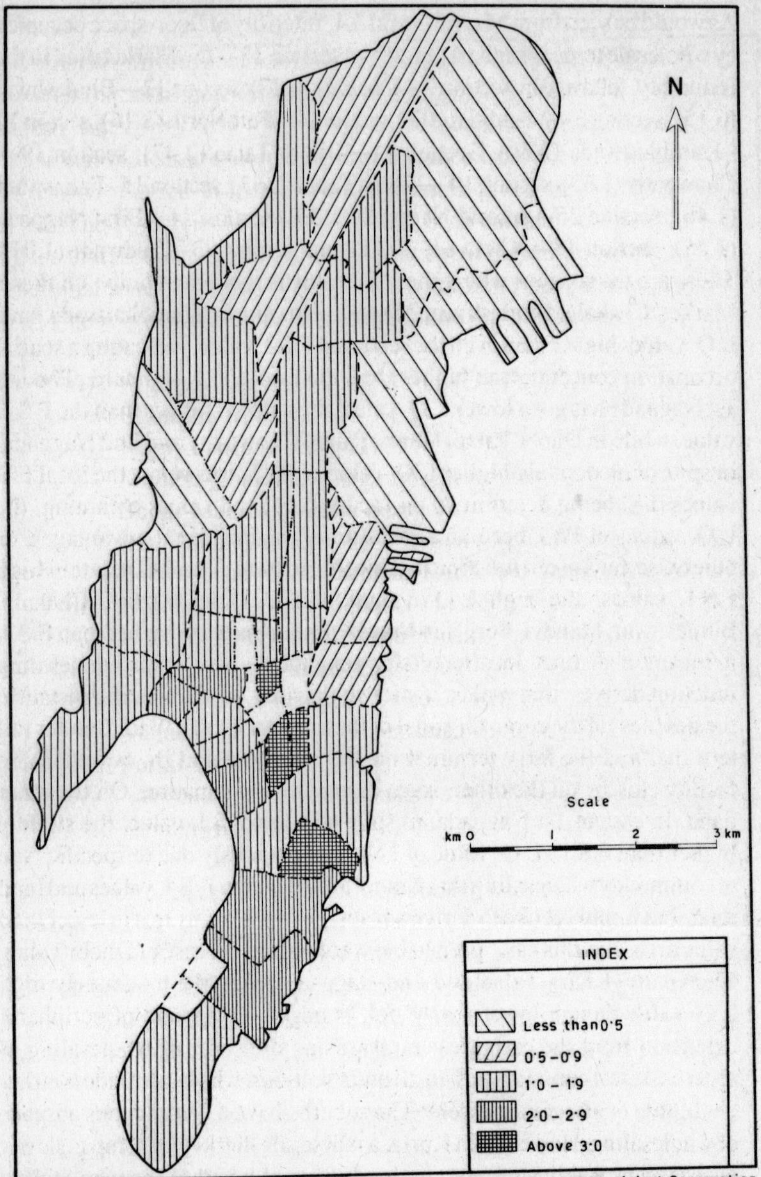

Source: Bombay Municipal Corporation.

MAP 5.2: Section-wise Floor-Space Index of the City of Bombay

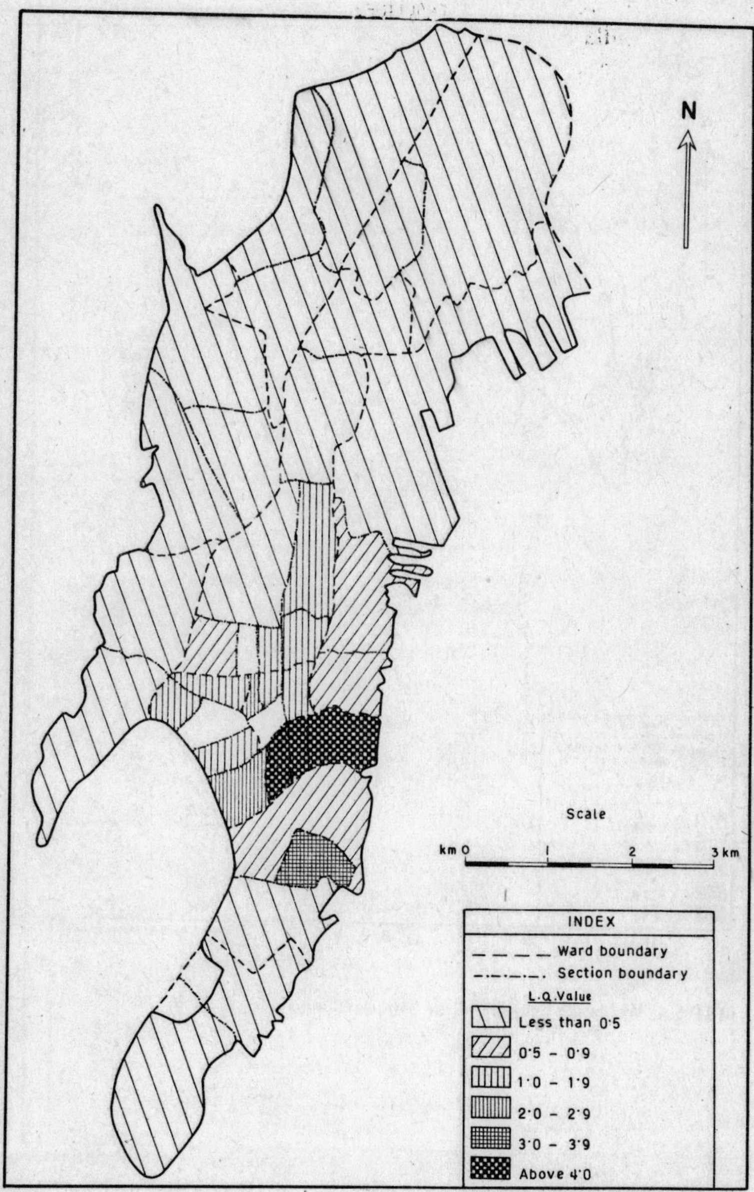

Source: Bombay Municipal Corporation.

MAP 5.3: Wholesale Intensity of the City (Bombay)

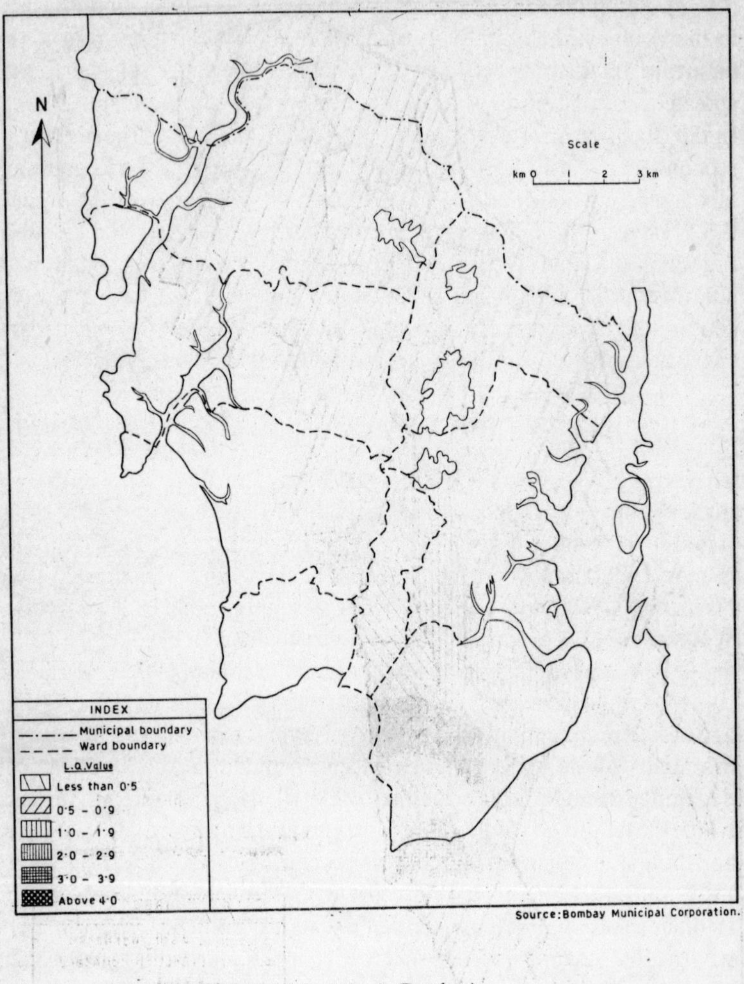

MAP 5.4: Wholesale Intensity of the Suburb (Bombay)

5.2.5 *Intensity of Retail Trade (IRT)*

Maps 5.5 and 5.6 show the intensity of floor space occupied by retail trade (I.R.T). Retail function, naturally, would be present throughout the city, in the neighbourhood of every sizeable residential area because of the nature of interaction between the functions. However, the locational spread or concentration of retail functions would vary from area to area. In the island city, for example, the L.Q. value of I.R.T. is the highest in section 4, Fort North (6.50), corresponding to the fifth highest concentration of wholesaling and fourth largest intensity of godown, with the highest F.S.I. values (3.81). This area has a rather low intensity of residential function (0.5) but is the hub of day-time activity associated with administration and business functions and with the rail and road commuting, service-seeking population converging on its roads on way to the administrative offices, banking houses and business houses.

5.2.6 *Intensity of Residential Function (IRF)*

The intensity of residential function (I.R.F) based on floor-space utilisation in the sections/wards were also calculated and plotted in Maps 5.7 and 5.8. It is seen from the maps that in the island city, the L.Q. values of the IRF is most sections have values little more than unity. Section 25, 2nd Nagpada, has the highest score (1.50) closely followed by section 20—Walkeshwar (1.40). The 10 other sections are number 31, 32, 21, 15, 17, 18, 33, 16, 10 and 14 in a descending order of ranking where the I.R.F. values are more than unity. In section 8, the I.R.F. value is unity. As field and observational experience shows, these are the main residential areas, essentially of the upper middle and lower middle income groups. The declining residential core of Kumharwada, Bhuleshwar and Kamathipura where the retailing - wholesaling functions are usurping residential areas, and the textile mill industrial core of the central part and their fringes have I.R.F. values below unity but close to it. In these areas, residential function is of a lesser importance. In the remaining sections, i.e., the Fort, wholesaling complex, Worli and Chinchpokli, residential function is even less. The suburbs show a markedly high concentration of residential function higher than even the highest I.R.F. values in the city. The I.R.F. values of wards R. (1.90), H. (1.80), K. (1.70), M. (1.70) and P (1.60) are close to 2.00, while wards L (1.40), T (1.40) and N (1.20) have I.R.F. values more than unity. The western suburbs especially to the west of the railway are pure residential areas. Only Chembur (M) of the eastern suburb has a higher residential function.

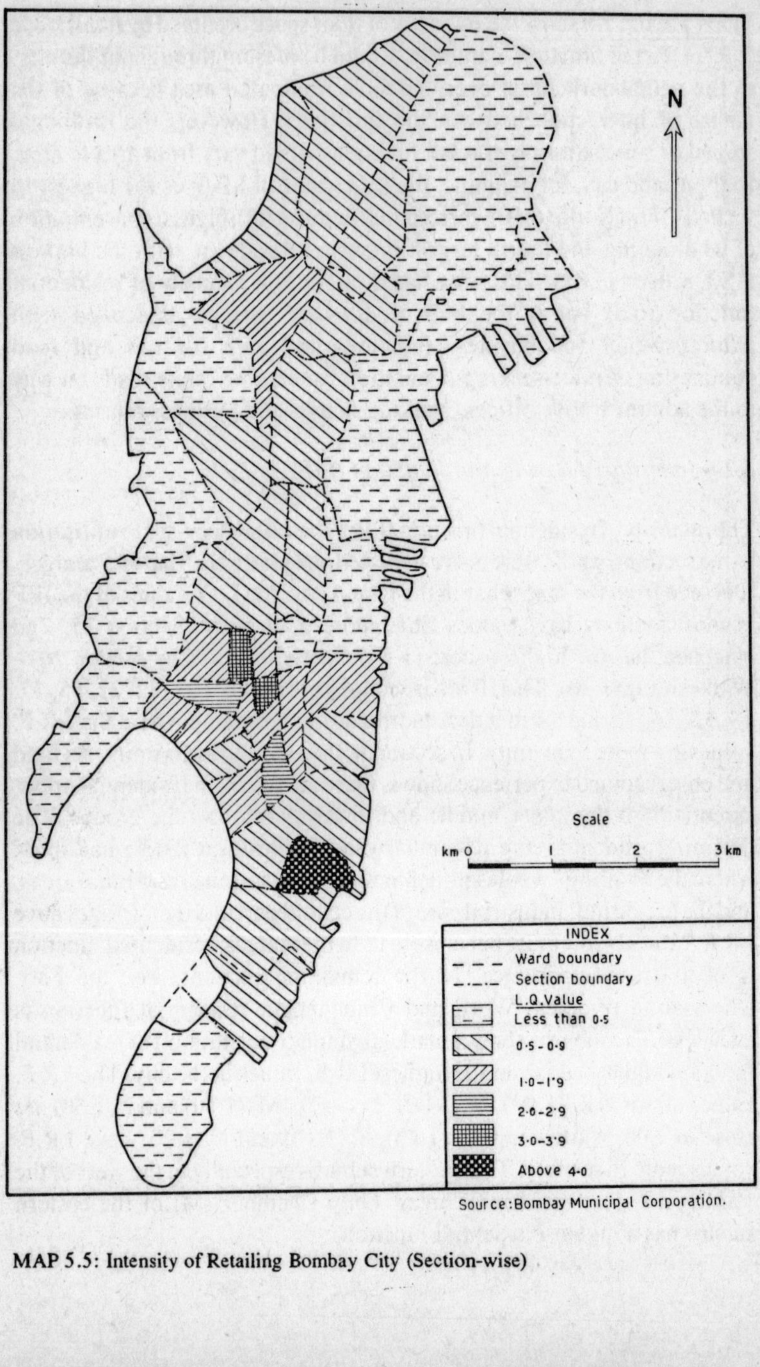

MAP 5.5: Intensity of Retailing Bombay City (Section-wise)

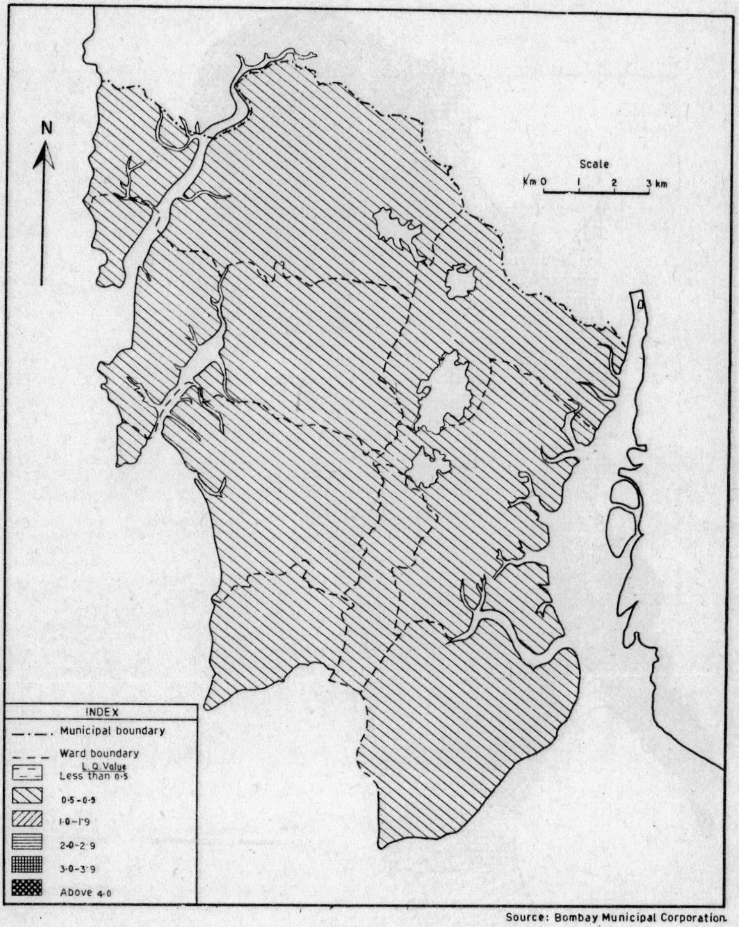

MAP 5.6: Intensity of Retailing Bombay Suburb (Ward-wise)

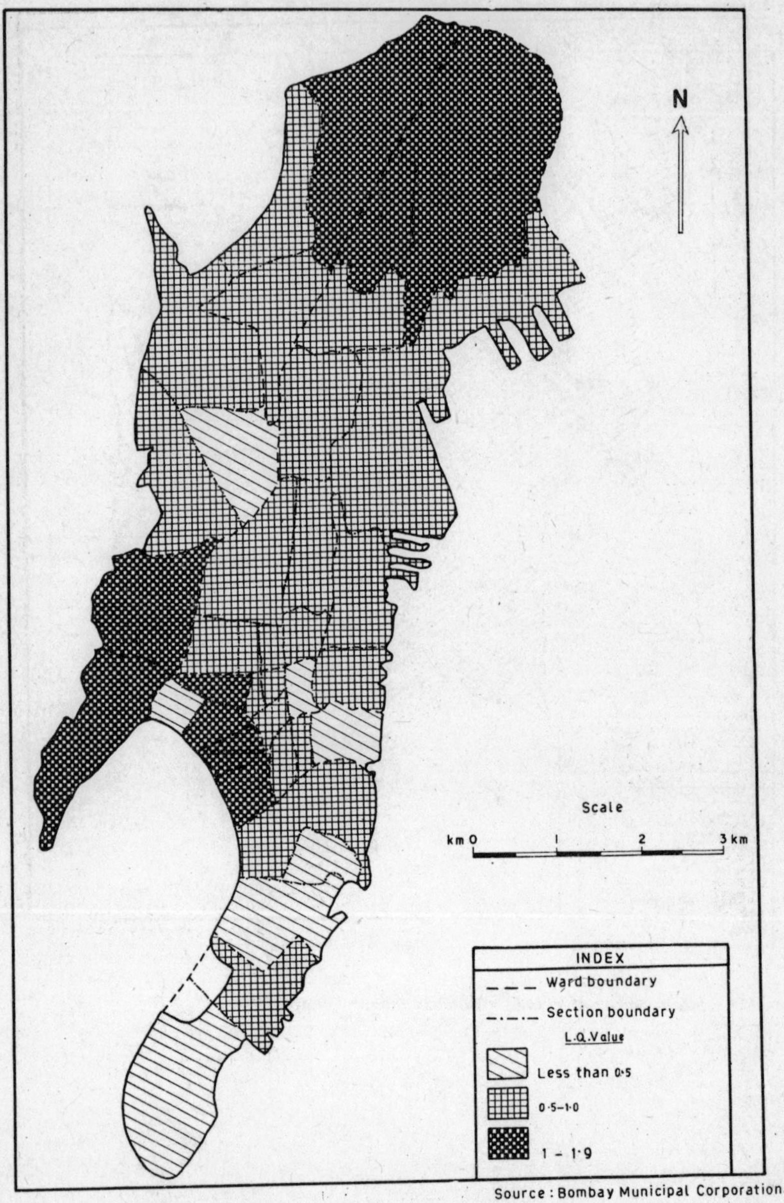

MAP 5.7: Intensity of Residential Function–Bombay City (Section-wise)

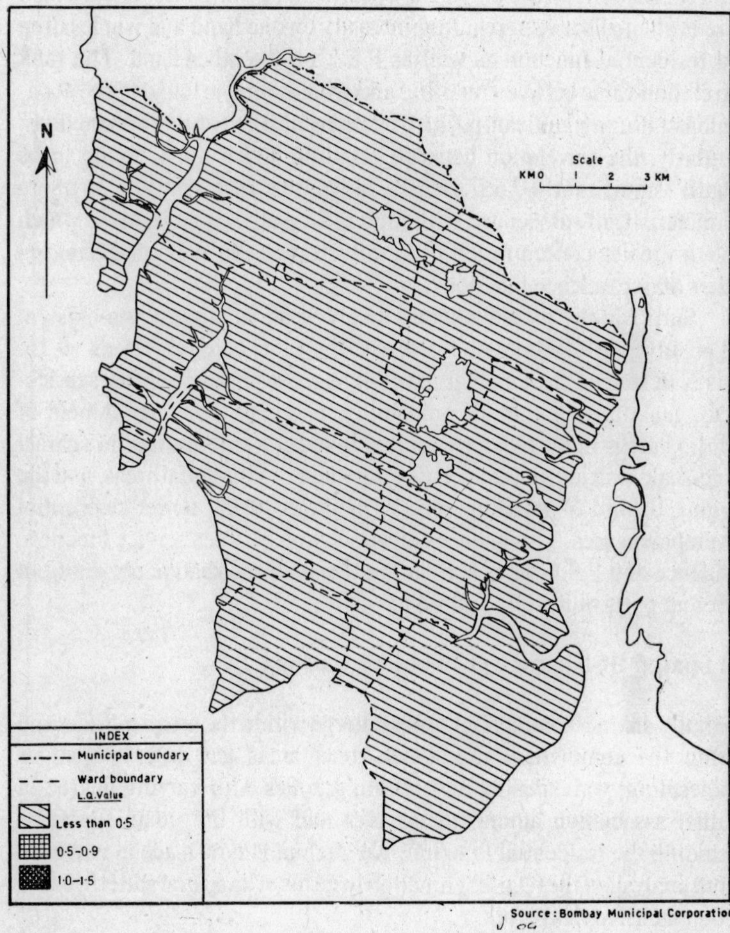

MAP 5.8: Intensity of Residential Function—Bombay Suburbs (Ward-wise)

5.3 Nature of Association between F.S.I., Wholesaling, Retailing and Residential Functions

Rank correlation co-efficients using Spearman's's[3] Rank Correlation method were evaluated between retailing intensity on one hand and wholesaling and residential function as well as F.S.I. on the other hand. The rank correlation value between retailing and wholesaling is found to be +0.61, demonstrating a significant positive relationship between the two functions. Similarly, the correlation between retailing and F.S.I. is found to be equally significant +0.66, clearly indicating that retailing is more characteristically associated with high F.S.I. values, that is, areas which have a greater concentration of administration, business and banking rather than residence functions.

Surprisingly, retailing shows a very weak, an almost non-existent but positive correlation with residence, the value being as low as +0.16. This is in spite of the fact that retailing is a consumer-oriented, service-sector function. This, in all probability, is due to the entrenchment of retailing areas in the older areas close to the city core, its inertia against large-scale shifting, unplanned development of residential areas, and the lagging behind of retailing service facilities in the newer residential development area. This analysis of association between service function, residence and F.S.I., however, shows the areal association obtaining in different parts of the city and suburb.

5.4 Spatial Distribution of Retailing Clusters

Spatially distinct commercial zones emerge within the urban mosaic, and within the commercial functional zone, areas get differentiated as wholesaling, warehousing and retailing zones with varying degree of mutual association among themselves and with the urban functions including the residential function. An attempt is now made to make in-depth analysis of the retailing function in terms of its spatial pattern, based upon extensive field-work.

5.5 Retail Cluster Zones

Based on experience and knowledge gathered in the field, and observed retailing consumer affinities and preferences, Greater Bombay region in this study was grouped into 22 zones (as shown in Table 5.2 and Map 5.9 keeping in mind the socio-economic homogeneity and existing retailing affinities. They are, however, adjusted to broadly conform to section boundaries for practical reasons. Detailed street - to - street field - work

TABLE 5.2: Retailing Zones of Greater Bombay

	Name of the Zones	Broad Sections of Metropolitan Corporation of Greater Bombay
1.	Colaba	1, 2
2.	Fort	3, 4, 5
3.	Mandvi-Chakala	6, 7
4.	North and South Dock Area	9, 22, 29
5.	Bhuleshwar-Market	8, 10, 11, 12, 13
6.	Girgaum-Chowpatty	14, 15, 16, 18, 19
7.	Walkeshwar-Malabar and Cumbala Hill	20, 21
8.	Tardeo-Byculla	23, 24, 25, 26, 17, 27
9.	Worli-Chinchpokli	36, 37, 38
10.	Dadar-Mahim	33, 34, 35
11.	Parel-Naigaon-Wadala	28, 30
12.	Matunga-Sion	31, 32
13.	Bandra-Pali-Khar-Santacruz (W) Danda	Parts of 43, 44, 39, 40, 41, 42, 46, 47
14.	Bandra Colony-Santacruz-Kalina	Parts of 43, 45, 48
15.	Andheri (W)-Juhu-Irla-Jogeshwari (W)	50, 51, 52, 53, 54, 57
16.	Vile Parle-Andheri-Jogeshwari (E)	49, 55, 56
17.	Malad (W)-Erangal-Goregaon (W)	60, 61, 62, 63, 65
18.	Goregaon (E)-Aarey-Malad (E), Dindoshi	58, 59, 63, 64
19.	Kurla-Chembur-Trombay	75, 76, 77, 78, 79, 80
20.	Ghatkopar-Vikhroli	8, 82, 83, 84, 85, 86, 87, 88
21.	Borivli (W)-Manori-Gorai	66, 67, 69, 70, 71
22.	Borivli (E)-Poisar-Akurli, Dahisar (E)	68, 72, 73, 74

"E" and "W" after the names of the zones indicate East and West, respectively.

was done by moving on foot in all parts of the city and suburbs where retailing shops have developed in clusters, and enumerating, identifying and field mapping of the shopping characteristics in each of the areas. The clusters were identified on the basis of presence of an alignment of continuous shopping front along one or more main roads and the adjoining by-lanes to present the appearance of a shopping area. Though no definite or specific minimum number of shops was used as a delimiting criterion, the presence of at least 50 shops in the suburbs and 100 in the city was also used as an additional subjective qualifying condition.

Retailing affinities were decided upon in terms of (i) accessibility of the cluster within pedestrian walking distance or through various vehicles

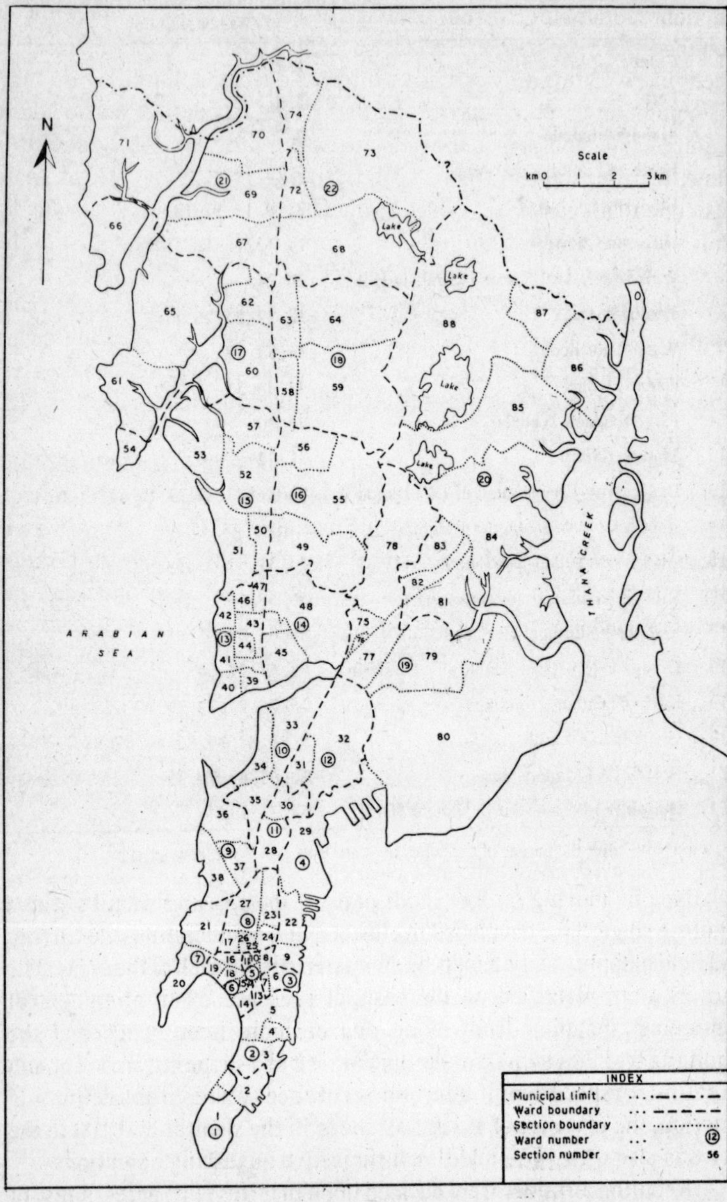

MAP 5.9: Zoning of the Study Area (Greater Bombay)

of public conveyance, i.e., within a distance of 3 to 5 kms, or within half an hour commuting distance, and (ii) the range of availability of commodities such that, except for specialised and specific requirement needs like wedding dresses and jewellery or interior decoration furniture, all commodities necessary for normal living, including that of daily requirements, and seasonal needs are procurable within the zone. It is quite likely and possible that within a retailing zone, so identified, more than one retail cluster may exist, which between themselves practically satisfy all the needs of normal living of the consumer population of the zone.

During field-work in different parts of wards/sections for identifying retail cluster, a physical enumeration of the shops in each of the clusters was also made and they were differentiated according to commodities offered for retailing. The clusters identified in the field have been plotted in Maps 5.10 and 5.11.

The number of shops in the shopping clusters so identified through field-work differ from the number as identified by shops and establishments unit of B.M.C. This is due to the difference in the classificatory system adopted. A variety of services identified as shops by B.M.C. for taxation purposes which do not offer any product for the consumers are not included as shops in this study—as, for example, hair dressing saloons, tailoring shops, eating houses—though they also take advantage of the consumer population congregating in the clusters for shopping, to gather a clientale.

5.6 Spatial Variations in Retail Clustering Tendency within Greater Bombay

The shops so identified in clustered groups number 16,399 and these were grouped into 22 zones on the basis discussed above. Table 5.3 indicates the number of shops enumerated in the 22 zones, zone-wise. The shops and establishments unit of B.M.C. has identified 1,30,699 shops for its purpose. All of them are not shops in the true sense of the term; also all of them do not occur within the cluster areas. Field experience shows one in three of the shops enumerated by the B.M.C. in the retail cluster areas are not shops in the true sense but are service establishments. Secondly, all shops do not occur in the cluster areas alone. Many street corner shops in poor residential areas cater to the immediate local needs like vegetables, fruits, general grocery, etc. A broad comparison of the shops in clusters identified in 22 zones with the number of shops as per shops and establishments department of B.M.C. shows that there is a strikingly

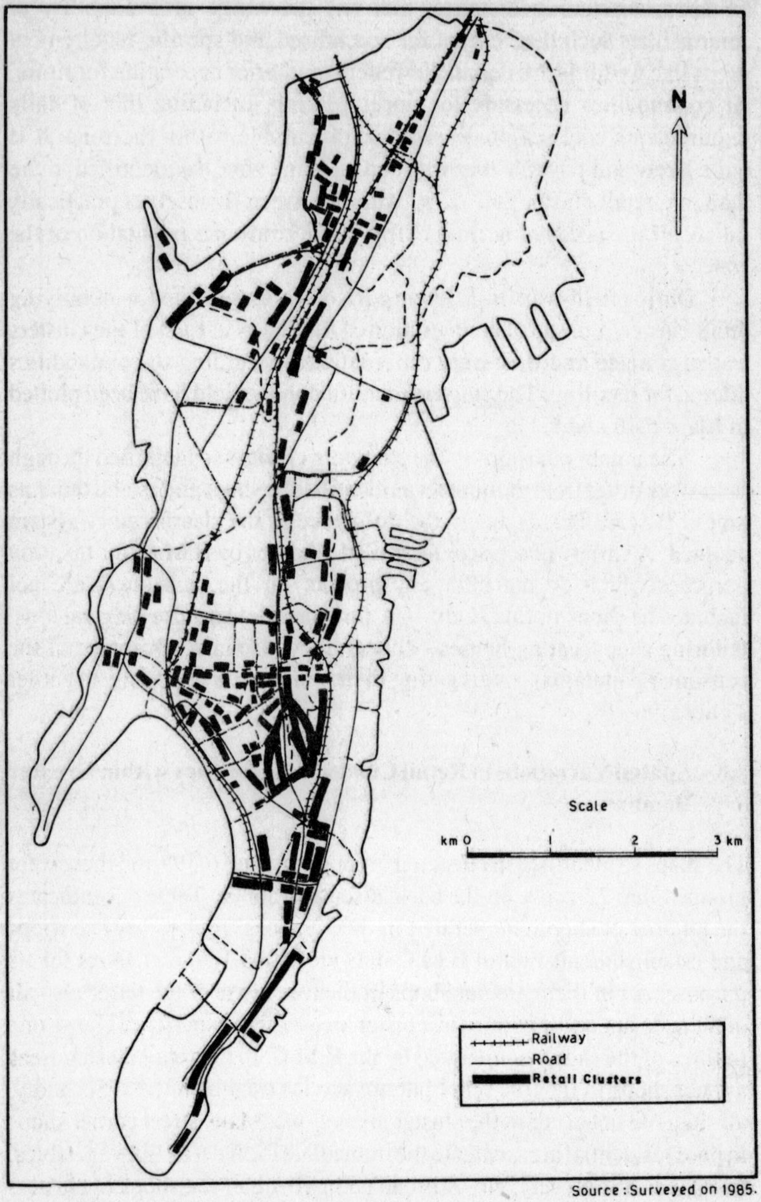

MAP 5.10: Retail Land-Use of Bombay City

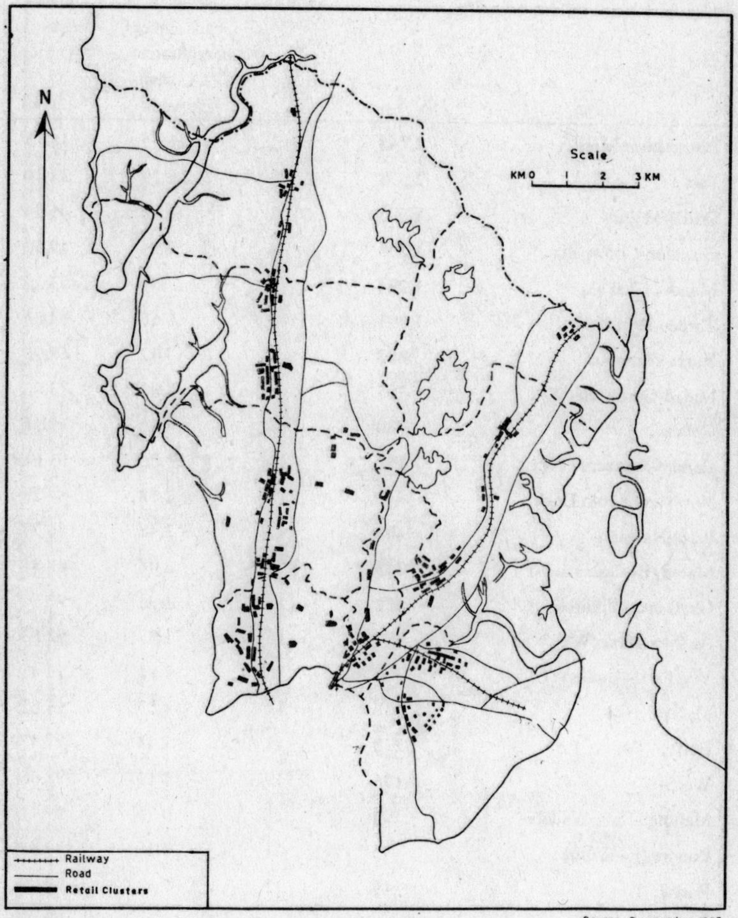

MAP 5.11: Retail Land-Use of Bombay Suburb

TABLE 5.3: Cumulative Percentage of Retail Establishments in Clusters

Name of the Zone	No. of Retail Establish-ments	Rank as per No. of Establi-shments	Percentage of Retail Establish-ments in each Zone in relation to Total Estab-lishments	Cumulative Percentage of Retail Establish-ments
Bhuleshwar Market	2,748	1	16.75	16.75
Fort	2,673	2	16.29	33.04
Dadar-Mahim	1,395	3	8.50	41.54
Girgaum-Chowpatty	1,306	4	7.96	49.50
Mandvi-Chakala	1,294	5	7.89	57.39
Tardeo-Byculla	1,091	6	6.65	64.04
Kurla-Chembur	807	7	4.92	68.96
Malad-Goregaon (W)	795	8	4.84	73.80
Colaba	730	9	4.45	78.25
Bandra-Santacruz (W)	568	10	3.36	81.61
North and South Dock	440	11	2.68	84.29
Parel-Naigaon	408	12	2.48	86.77
Malad (E)-Goregaon (E)	335	13	2.04	88.81
Ghatkopar-Vikhroli	321	14	1.95	90.76
Andheri-Juhu (W)	308	15	1.87	92.63
Vile Parle-Andheri (E)	300	16	1.82	94.45
Malabar Hill	258	17	1.57	96.02
Borivli (W)	238	18	1.45	97.47
Worli-	196	19	1.19	98.66
Matunga-Sion-Wadala	124	20	0.75	99.41
Bandra (E)-Kalina	45	21	0.27	99.68
Borivli (E)	19	22	0.11	99.99

significant difference in the shopping groupings between the city and suburbs, older residential areas and newer residential areas. The older residential core as well as the commercial zone, and the CBD area have bulk of the retailing functions concentrated in well-organised retail clusters, while in the industrial and low-income areas of Central Bombay, retail clusters are not so well-developed, considerable retailing being carried on in individual isolated shops. In the suburban zone, away from the rail head retail clusters, shopping is totally dispersed, especially so in the residential areas away from the railway. The bulk of the retailing shops in the suburbs are in such dispersed and scattered locations, rail-head clusters accounting for only a small proportion of the total number of shops. This appears to be due to the time-lag in the actual retail cluster development following other infrastructural development and a minimum level of occupation of the residential development areas in the newly developing suburban areas, while in the city proper maturity in the phase of retail organisation is indicated in the development or otherwise of retail clusters.

As discussed above, all the retailing clusters in the city and its western and eastern suburbs were identified and mapped as shown in Maps 5.10 and 5.11. The maps show distinctly a heavy clustering of retailing in the CBD and the adjoining wholesaling zone to its north, with a linear sprawl along the main roads, northwards and also south into Colaba Causeway area (Shahid Bhagat Singh Road). A secondary and minor concentration, though of a much lesser scale, is noted around Dadar with the convergence of both central and western rail commuting and a substantial crossing over of population from either side.

In the suburbs, commuting rail-heads have almost an exclusive siting of the shopping areas, except for the significant exceptions in Chembur Naka and Bandra Linking Road. Away from the rail-head, shopping clusters abruptly peter off, in suburbs.

Based on the distribution pattern of the retail clusters, road linkages, commuting rail-heads and field evaluation of consumer behaviour and preferences for shopping areas, twenty-two retailing zone groupings have been identified as shown in Maps 5.10 and 5.11 and discussed as under. It is pertinent to point out that these zones tend to be smaller in area in the older congested and better developed city centre, while along the peripheral suburban zones, where residential function growth is a matter of the last two decades, and is still an on-going process, the retailing zones are much larger in area partly due to differentials in rent values and partly due to a closer packing of shopping frontage and vertical development nearer the city core. The different retailing zones also vary in the support of the population they serve, but, on an average, support the service needs of 3 to 6 lakhs of people.

5.7 Retailing Spread Characteristics

It may be pertinently noted that certain distinctive features may be noted in the spatial pattern of organisation of retailing clusters in Bombay and its suburbs. The retailing zones 3,4 and 6 form an almost spatially contiguous commercial landscape with an areal spread all over, of the retailing function. However, from east to west, from Masjid, Mandvi to Girgaum, there is a continuous gradation and gradient from predominantly wholesaling through semi-wholesaling to predominantly or totally retailing. Retailing occupies all available niches and street fronts in the wholesaling area of Mandvi - Chakala, while its spread is more continuous and all-pervading with increasing distance and away from the wholesaling core. What perhaps is strikingly observable is the co-existence of mutual intimacy of the related and interacting commercial functions like wholesaling, semi-wholesaling, retailing, warehousing and godowns, transport carriers of all types, brokers and commercial middleman—their siting being influenced by land rental values, shopping frontages and consumer density in movement along roads.

This areal pattern of spread of the retailing function gives way to a distinctly linear pattern along the main roads and the by-lanes with a greater convergence along road junctions and chowks southwards from the commercial core through the Fort area (primarily Dadabhoi Naoroji Road, Mahatma Gandhi Road) into Colaba Causeway of retail zones 1 and 2. There is a linear ribbon-type dense concentration of shopping chains on both sides of Colaba Causeway (presently called Shahid Bhagat Singh Road). The development of this retail concentration has an early history: with the inauguration of the Wellington Pier (Apollo Bandar) in 1825 and construction of Colaba Causeway in 1838, a small shopping cluster had been growing fast, first for catering to the needs of white population of the neighbourhood. With the expansion of the Apollo Bandar, the emergence of a number of hotels in Colaba and the replacement of white population by rich Indians especially the Parsi and Christian population, Colaba Causeway (Shahid Bhagat Singh Road) not only retained its old fame of being a fashionable retail enclave, but also developed it even further. The high quality of retailing in this area is indicated by the presence of shops catering quality ladies garments, all types of jewellery trinkets, shoe shops, antiques and gift materials of tourist value, etc. All the shops are located on the front side of the buildings with attractive showrooms, window displays, efficient and well-behaved salesmanship. This retail concentration, rather a shopping mal, caters not only to Indians, be they residents of Colaba, or Bombay and the Metropolitan region, or domestic tourists who congregate in the star hotels of the area. Here again, land rent differentials between the main roads and the by-lanes, ground floor frontage and upper floor of buildings, corner locations all influenced by commuting pedestrian density variations, bring about subtle variations in the nature and intensity of retailing function which will be discussed in a later chapter.

Northward of the commercial core, retailing also reveals strong affinities for a linear spread along the trunk road arteries like Mohammdally Road, Victoria Garden Road (J.J. Road), Vincent Road (Babasaheb Ambedkar Road) on one hand and Princess Street (Shyamaldas Gandhi Marg), Girgaum Road (Jagannath Shankar Sheth Road), Kalbadevi Road, Opera House (P.Parluskar Chowk), Lamington Road (Dr. Dadashaeb Bhadkamkar Marg), Charni Road (Raja Rammohan Roy Road), and Grant Road (Moulana Saukatali Road) on the other. These are the arteries of heavy road commuting by public conveyance. These linear spreading retail clusters develop nodes in the road chowks or circles like the Opera House (P. Parluskar Chowk), Nana Chowk or Worli Naka (Acharya Atre Chowk).

Opera House (P. Parluskar Chowk) is located between Girgaum Road (Jagannath Shankar Sheth Road) and Nana Chowk. It is a specialised retail complex having specialised goods: automobile and auto-spare parts, jewellery, electronics and electrical goods, radio, T.V. parts, quality garments etc. There is also a heavy concentration of service sector establishments like polyclinics, specialist-doctors and chartered accountants in this area.

A secondary, although of a much lower order than the main commercial complex of B and C wards, is the areally spread retailing cluster that has developed around Dadar-Naigaum Khodadad Circle (formerly known as Dadar T.T.). The convergence and transfer of commuting population of both Central and Western Railways at Dadar Junction and the east-west road flows across the Tilak Bridge of pedestrian population seem to explain this cluster which primarily caters to the lower middle-income population groups of the northern part of the city as well as the suburban population, because of a greater proximity. A limited development of wholesaling of consumer perishables and utility commodities is the latest addition in this zone.

Northward, beyond the Mahim and Sion Causeways in the suburban area, the commuting rail-heads have the retailing cluster development. Each rail-head has, on one or both sides, along the roads, converging on the rail-head, shopping frontages catering to the consumer needs of the local suburban commuters. They have a heavy commercial transaction during the evening hours of the day, associated with the return traffic of the commuters on their way home. The size of the retail zone, its commodity differentiation and specialisation appear to be intimately related to the size and income levels of the commuting population.

Converging at the rail-head the major suburban stations like Chembur, Ghatkopar, Santacruz, Vile Parle, Andheri and Goregaon have a larger development of retail shopping than other suburbs like Khar, Jogeshwari, Vidyavihar and Vikhroli.

The suburban interiors, away from the rail-heads, have developed some retailing clusters often at the bus terminals or along some major roads, purely to satisfy the ubiquitous daily needs of local residents. In the recently developed, planned suburban areas like Bangurnagar, retailing has developed in the form of shopping centres in one or two buildings.

Thus, four different types of retail clusters can be identified in Greater Bombay:

(1) The areally spread commercial zone in the city centre where the whole landscape is dominated by the commercial functions, retailing constituting a major or whole component.
(2) The linear pattern along the main arterial roads in the older residential zones of the city, with retail nodes in chowks/ circles.
(3) The commuting rail-head clusters in the suburban areas.
(4) Limited shopping complexes on the newer suburban planned development areas away from the rail-head.

5.8 Statistical Distribution of Retailing among the 22 Zones

The above descriptive analysis of the areal distribution of retail clusters in Greater Bombay in the 22 zones has been augmented by analysis and graphical presentation to give a measure of the concentration of retailing clusters. The retail zones were ranked in a descending order according to the areal density of retail shops of the zones.[4] Tables 5.3 shows the relevant cumulative data. A graph (Fig. 5.1) was drawn by plotting the cumulative percentage of retail establishments occurring in clusters against their rank orders.

It will be seen from the graph that a third of the retail establishements in clusters occur within the first and second ranking areas alone, namely, Bhuleshwar and Fort. Almost 50 per cent of the retail establishments occur within the first four ranking zones, namely, Bhuleshwar, Fort, Dadar-Mahim and Girgaum-Chowpatty. Only 20 per cent of the retail establishments in clusters occur in the bottom twelve ranking order zones.

5.8.1 *The Lorenz Curve*

A Lorenz Curve[5] was also drawn by plotting the cumulative per centage

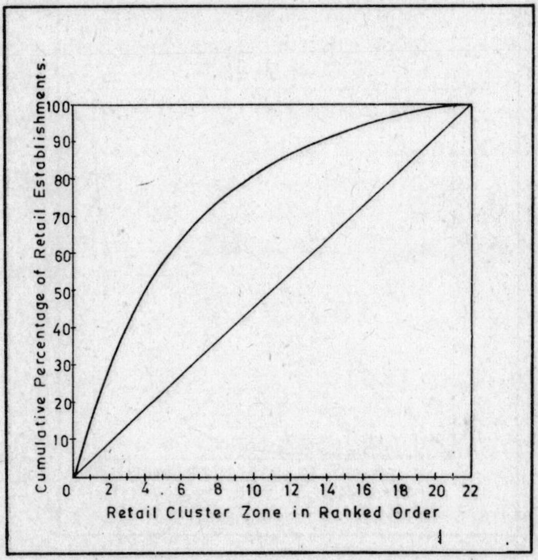

Fig. 5.1: Cumulative Curve of Retail Establishments

of retailing establishments in clusters against the cumulative percentage of corresponding areas in retail zone as shown in Fig 5.2 and Table 5.4. From this curve, it will be noted that 75 per cent of the retail establishments that occur in the retail clusters are accounted for by 10 per cent of the area. Thirty five per cent of the area has 8 per cent of shops. The curve clearly demostrates the high degree of concentration of retail clusters in very limited areas. The Lorenz Curve in this case is far away from the 450 line, namely, the line of even distribution.

5.8.2 *The Gini's Co-efficient*

While the area under the Lorenz Curve expressed as a per centage of the area under the 45° line will give graphically the degree of concentration which in this case is found to be (0.79), Gini's co-efficient, or the co-efficient of evenness, is a more precise quantitative measure of this concentration. It is found by using the formula

$$R = \frac{1}{100 \text{ X } 100} \quad (x_0 \, y_1 \; - x_1 \, y_2 - \ldots . \; x_n - y_n)$$

$$= (x_0 y_0 + x_2 y_1 + \ldots . + x_n y_n - 1)$$

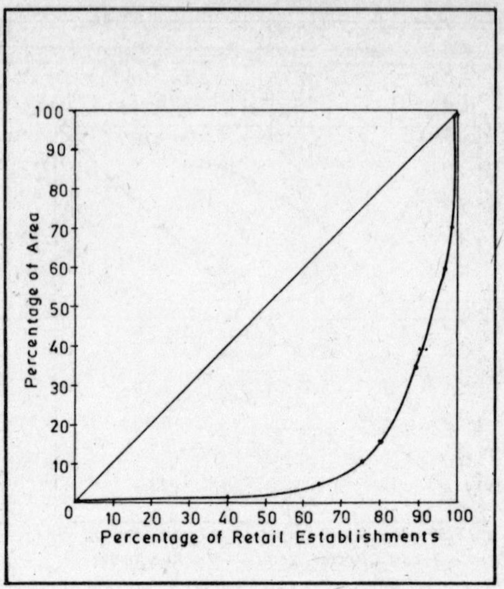

Fig. 5.2: Lorenz Curve showing the Relationship between the Retail Establishments and Area in each Zone

where x_0y_0 is the point of origin and x_1 - x_n is the cumulative per centage of corresponding areas, n is the number of zones, in this case 22. By substituting for the values of x_1 - x_n. y_1 - y_n in the formula, the Gini's co-efficient is found to be 0.77. Theoretically, the Gini's co-efficient or the co-efficient of evenness, which gives the summary measure of the extent of inequality of distribution, can vary from zero to unity depending upon the evenness of distribution. The value 0.77 for the co-efficient in this distribution indicates the high degree of inequality in the distribution of retail establishments in clusters, or in other wards, it is a measure of the degree of locational concentration.

5.9 Spatial Hierarchy of Retail Cluster Zones

While the above two graphs help in bringing out the clusters in Greater Bombay as a whole based upon the statistical ranking of the zones, they, however, do not take into consideration the spatial arrangement of the retail zones. In order to appreciate in a better manner the spatial ogranisation of retail clusters, keeping Fort North area as the city centre, the retail zones were arranged in a spatial order according to increasing

TABLE 5.4: Relation Between the Area and Number of Establishments in each Zone

Name of the Zone	No. of Establishments in each Zone	Area in each Zone	Density of Establishments	Ranking as per Density of Establishments	Cumulative Area (km^2)	Percentage of Area in each Zone relation to Total Area	No. of Cumulative establishments	Percentage of Establishments in each Zone in Relation to Total Establishments
Bhuleshwar	2748	.1.28	2147	1	1.28	0.3	2748	16.7
Mandvi	1294	0.87	1487	2	2.15	0.5	4042	24.6
Fort	2673	4.81	556	3	6.96	1.6	6715	41.0
Girgaum	1306.	2.38	549	4	9.34	2.2	8021	48.9
Tardeo	1091	5.18	211	5	14.52	3.4	9112	55.5
Dadar	1396	9.38	149	6	23.90	5.6	10508	64.0
Colaba	730	5.65	129	7	29.55	6.9	11238	68.5
Parel/ Naigaum	408	3.98	102	8	33.53	7.8	11646	71.02
Bandra/ Santacruz	568	9.03	63	9	42.56	9.9	12214	74.5
North and South Dock	440	7.96	55	10	50.52	11.7	12654	77.1
Malabar Hill	256	5.11	50	11	55.63	12.9	12910	78.7
Worli	196	8.50	23	12	64.13	14.9	13106	80.0
Malad/ Goregaon (W)	795	37.86	21	13	101.99	23.72	13901	84.7
Andheri (W)	308	23.64	13.0	14	125.63	29.2	14209	86.6
Malad/ Goregaon (E)	335	25.81	12.97'	15	151.44	35.2	14544	88.7
Vile Parle/ Andheri (E)	300	23.59	12.7	16	175.03	40.7	14844	90.5
Kurla/Chembur	807	67.58	12.0	17	242.61	56.4	15651	95.4
Matunga/Sion/ Wadala	124	12.60	10.0	18	255.21	59.4	15775	96.2
Borivli (W)	238	36.17	6.6	19	291.38	67.8	16013	97.6
Bandra (E)/ Kalina	45	10.50	4.3	20	301.88	70.2	16058	98.0
Ghatkopar/ Vikhroli	321	87.54	4.0	21	389.42	90.6	16379	99.8
Borivli (E)	19	40.47	0.5	22	429.89	100.0	16398	100.0

TABLE 5.5: Relation Between Distance from the City Centre and Percentage of Retail Establishments in Clusters

Sr. No.	Area	Average Distance from the City Centre	Percentage of Retail Establishments in Clusters in Relation to Total Establishments in Greater Bombay
1.	Fort	0.00	16.29
2.	Mandvi-Chakala	1.50	7.89
3.	Colaba	2.00	4.45
4.	Bhuleshwar-Market	2.21	16.75
5.	Girgaum-Chowpatty	2.21	7.96
6.	Malabar Hill	3.11	1.57
7.	Tardeo-Byculla	3.50	6.65
8.	Worli-Chinchpokli	5.50	1.20
9.	North and South Dock	6.10	2.68
10.	Parel-Naigaon	7.65	2.49
11.	Dadar-Mahim	9.00	8.51
12.	Matunga-Sion	10.22	0.75
13.	Bandra-Santacruz (W)	16.22	3.36
14.	Bandra (E)-Kalina	16.24	0.27
15.	Kurla-Chembur, Trombay	16.51	4.92
16.	Andheri (W)-Juhu-Irla	21.50	1.87
17.	Vile Parle, Andheri (E), Jogehwari (E)	21.50	1.82
18.	Ghatkopar-Vikhroli	22.50	1.96
19.	Goregaon (E)-Malad (E), Dindoshi	25.90	2.04
20.	Malad (W)-Erangal-Goregaon (W)	26.50	4.84
21.	Borivli (E), Poisar	33.05	0.11
22.	Borivli (W), Manori, Gorai	34.03	1.45

distance from the city centre as shown in Table 5.5. A histogram (Fig 5.3) was plotted showing the relative per centage of retail establishments in clusters against their spatial order with an increasing distance from city centre.

The following major conclusions may be drawn from the plot shown

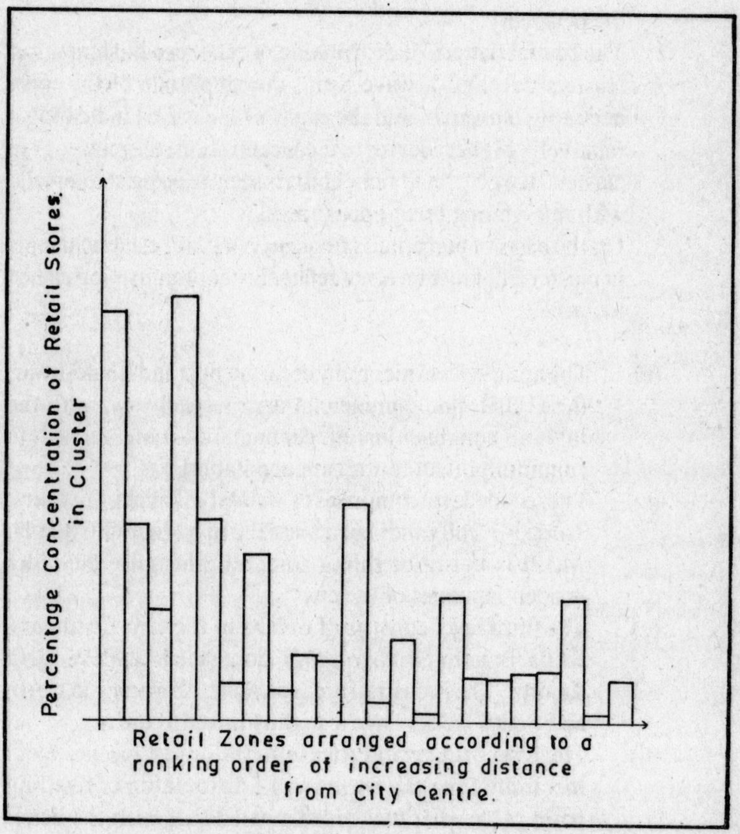

FIG. 5.3: Bar Graph showing the relationship between the Average Distance from the City Centre and the Percentage of Retail Establishments in Clusters

in the diagram:

1. Broadly, there appears to be a distance decay of clustered retailing from city centre outwards towards the suburbs and the periphery.
2. The spatial pattern of distribution of retail establishments in clusters develops a wave form, the amplitude of the wave declining outwards, and the crests of the waves indicating a relatively higher degree of concentration of retailing in clusters. Larger "retail rich" clusters seem to occur at intervals with intervening "retail poor" areas.
3. On the basis of percentage frequency of retail establishments in clusters, distinct levels of retail cluster organisation can be identified:

(a) The highest level hierarchy occurs in Fort and Bhuleshwar, the C.B.D. and commercial area respectively, with the highest congregation of commuters—more related to commuting than to resident population.

(b) The second level comprises of Mandvi-Chakala, Girgaum, Tardeo-Byculla and Dadar areas Barring Mandvi-Chakala, which is the wholesaling core, all others are the older residential zones of the city.

(c) The third level consists of Colaba in the city, Chembur - Kurla, Bandra-Santacruz (W), Goregaon Malad (W) all in suburbs. The industrial areas of Parel, Naigaon and North and South Docks have a slightly lower frequency.

(d) The rest of the retail zones form the lowest level of retail hierarchy from the frequency of distribution of retailing point of view. It may also be noted that even the better developed suburban retailing clusters are of a much lower order of development, indicative of their retailing youthful character. It is pertinent to point out that upper class residential area, like Malabar Hill, Cumbala Hill, Worli, Juhu have a low level of development of retail clusters - private conveyance facilitating their movement over distance to shopping areas of their choice.

REFERENCES

1. Nissel, H. (1977): "Greater Bombay—An Explanation of Structure, Function and Development" *Bombay Geographical Magazine,* Bombay - Vol. XXV, No. 1, pp. 1 - 27.
2. Analysis of the Functional Character of the Suburbs has largely been taken from:
   Arunachalam B. (1985): "Bombay", Machael Pacione (ed.), *Planning for Third World Cities,* pp. 121—125.
3. Formula for computing Rank Correlation has been taken from:
   Cole, J.P. and King, C.A.M. (1969): *Quantitative Geography,* University Press, Glasgow, p. 136.
4. It may be argued that the density could have been worked out against 'population' as against 'areas'. It was realised that the retail shopping clusters respond more closely to the commuting pedestrian flows that pass through the cluster area rather than the resident population within the boundary limits of the zone; hence, resident population was not likely to be a useful measure; therefore, 'area' was chosen.
5. Cole, J.P. and King, C.A.M. (1969) *Quantitative Geography,* University Press, Glasgow, p. 243.

6

FUNCTIONAL HIERARCHY OF RETAIL CLUSTERS IN GREATER BOMBAY

6.1 Introduction

While reviewing the literature on the subject of inter- and intra-urban hierarchy (in Chapter II), it has been pointed out that in developing countries not many relevant studies have been conducted. As for India, three important studies in the area of intra-urban hierarchy are mentioned in the review of literature, viz., the studies of Berry and Rees (1969)[1] and Dutta (1969)[2] in regard to Calcutta and Prakasa Rao (1983)[3] in regard to Bangalore. Hitherto, there has been no attempt towards determining the retail hierarchy of Bombay, the 'Urbs Prima in Indis'.

In this chapter, an attempt is made to establish a retail functional hierarchy with the help of appropriate quantitative techniques and established models already developed by urban geographers and to relate the same with a variety of retailing and other urban parameters, apart from the socio-economic zoning obtaining in Greater Bombay.

Based on the result of the intra-urban retail hierarchy, six areas with predominant retail functions are selected for subsequent in-depth case study of the retailing system, with the help of detailed field-work.

6.1.1 *Data Base for Retail Functional Hierarchy of Bombay*

The limitation of the B.M.C. secondary data has already been pointed out in an earlier chapter. There are 1,35,545 registered shops in Greater Bombay, as per the B.M.C. records.[4] These include both wholesale and retail establishments, the break-up of data in terms of wholesale, semi-wholesale and retail establishments is not available from the records of B.M.C. But, it is quite obvious that retail establishments outnumber wholesale establishments by many times. Also data on retail establishments based on types of commodities sold (i.e., number of retail functions) are not officially available from the records of B.M.C. It is for this reason that a field survey of 16,398 retail shops selling commodities (and not services) occurring in clusters in different parts of Bombay was undertaken.

6.1.2 *Qualitative Hierarchic Grouping of Retailing Zones*

The 16,398 enumerated shops are 12 per cent of the officially registered (with B.M.C.) wholesale and retail shops and service establishments. In reality, the number of retail establishments enumerated in clusters will take a far larger share of the retail shops alone though it is difficult to estimate the percentage of enumeration of shops in the clusters to the total.

In the last chapter, it was brought out that in terms of spatial frequency of retail shopping in clusters, some areas stand out as areal, linear and rail-head clusters. However, commodity grouping or areal specialisation as obtaining in different clusters was not indicated. Hence, in this chapter, the 16,398 shops enumerated have been grouped according to commodity types. There are literally thousands of commodities traded in a large metropolis like Bombay. It is impossible to handle these huge varieties of commodities without properly grouping them. When the field data were processed, a group of 26 commodity classes was constructed (as shown in Table 6.1) taking into consideration the commodity affinities and commodity association: for example, textiles include cotton, silk, wool and synthetic cloth variety. It is also observed that readymade garments and knit-wear tend to occur in association with textile shops and are often located side by side. This association develops because of their mutual affinity that can be taken advantage of, by reducing consumer travel distance. The broad scrutiny of Table 6.1 reveals the following salient features:

1. All commodity groups are not well-defined and homogeneous while some like textiles, electrical goods, utensils, crockery, jewellery, medicine, perfumes, grocery etc are clearly defined as commodity classes. There are a few which tend to be less precisely defined like, for example, plastics, sports goods and toys, variety goods, tobacco products dry fruits, etc. These commodities have been grouped together based upon affinities. It may also be pointed out that fish and meat products do not

TABLE 6.1: Ranking Order of Retail Cluster Zones as Per Commodity Functions

Group No.	Commodity Group	No. of Zones having commodity functions	Rank	Hierarchic Level
1.	Electrical/Electronics	22	1	
2.	Grocery/Gen. Provisions	21	2	I
3.	Textiles	19	3	
4.	Sweetmeat	18	4	
5.	Footwear	17	5	
6.	Spare Parts/Machinery/Hardware	16	6	
7.	Utensils (copper, brass, stainless steel, aluminium etc.)	15	7	
8.	Tobacco Products	15	8	
9.	Variety Goods	14	9	
10.	Medicine	14	10	
11.	Paper	13	11	II
12.	Jewellery	13	12	
13.	Sports Goods, Toys	13	13	
14.	Crockery	12	14	
15.	Cold Storage	12	15	
16.	Vegetables	11	16	
17.	Sewing Materials	10	17	
18.	Watch, Clock	10	18	
19.	Books and Periodicals	9	19	
20.	Photography and Xeroxing	9	20	III
21.	Paint and Chemicals	8	21	
22.	Timber	8	22	
23.	Furniture	7	23.5	
24	Bakery-Confectionery	7	23.5	
25.	Perfumes	3	25	
26.	Jari-Glass-Miscellaneous Special Products	3	26	IV

Source: The table has been constructed on the basis of field investigation in 1985.

(Appendix II shows the detailed break-up of commodities.)

form part of this grouping since they occur exclusively in fish and meat markets.

2. Based upon the nature of occurrence of shops of these various commodities, the 26 commodity classes can broadly be identified as four major categories.[5]

(a) Ubiquitous or near-ubiquitous, that tend to occur in over 75 per cent of retail zones; these products include the top ranking five groups in Table 6.1

(b) Commodities which are fairly well-spread and distributed in 11 to 16 zones of retail cluster zones and are tending towards ubiquitousness. Eleven commodities belong to this level.

(c) The third level includes the commodity groups which are fairly concentrated in 7 to 10 retail zones, i.e., less than half the number of retail zones; they include eight classes.

(d) The last group of two commodities have a heavy concentration, and are not found among many clusters.

It may also be pointed out that some commodity classes reveal strong mutual affinities of zonal occurrences with almost one-to-one correspondence, as, for example, furniture and timber, grocery and sweetmeat, sports goods, toys and variety goods.

It is also to be noted that no retail cluster has all the 26 classes of commodities. Also, the number of commodity groups occurring in the different retail cluster zones varies considerably from a handful to a large number, thus facilitating identification of hierarchy groups based upon the number of commodities.

Spatially, the four hierarchy levels of commodity grouping form the following areal association, as shown in Table 6.2.

Hierarchy level I has 20 and above commodity functions, out of 26. It covers the following retail cluster zones:

Bhuleshwar-Market
Vile Parle-Andheri-Jogeshwari (E)
Fort,
Bandra-Khar-Santacruz (W) and
Chembur-Kurla.
(In all 5 zones.)

TABLE 6.2: Zone-wise Distribution of Commodities

Sr. No.	Zones	Electrical Goods	Grocery-Provisions	Textiles	Sweet-Meat	Footwear-Leather Goods	Machinery Utensiles	Tobacco and Products	Variety Goods	Medicine-Drugs	Paper, Printing & Stationery	Jewellery	Sports Goods & Toys	Crockery-Porcelain	Cold Storage	Dry Fruits Gen. Stores	Vegetables. Perishables	Sewing & Embroidary Materials	Watch, Clock, Pen & Gift Centres	Books & Peridoicals	Photography & Xeroxing	Paints & Chemicals	Timber	Furniture	Bakery & Confectionery	Perfumes	Jari, Glass & Miscellaneous	Total No. of Retail Establishments	No. of Commodity Groups present in Each Zone.
		1	2	3	4	5	6	7	8	9	10	11	12	13	14	15	16	17	18	19	20	21	22	23	24	25	26		
1.	Colaba	15	160	159	10	63	20	—	20	80	—	—	10	—	45	20	65	25	—	15	8	—	—	—	15	—	—	730	16
2.	Fort	190	180	790	70	25	279	30	30	30	—	310	—	—	124	120	70	145	85	40	105	15	—	—	25	—	10	2673	20
3.	Mandvi-Chakla	110	500	30	—	—	—	—	—	—	10	58	—	—	40	30	—	—	10	—	—	370	16	10	40	—	70	1294	13
4.	North & South Dock	15	30	—	—	20	180	—	50	—	—	—	—	—	—	—	—	—	—	—	—	—	145	—	—	—	—	440	6
5.	Bhuleshwar-Market	205	46	830	95	223	166	153	—	10	145	170	205	75	40	40	90	-	30	30	10	45	—	5	—	25	110	2748	21
6.	Girgaum-Chowpatty	100	60	391	40	30	260	130	—	45	—	85	60	40	—	50	—	—	—	—	—	—	—	—	15	—	—	1306	13
7.	Malabar-Cumbala Hill	20	10	75	7	7	10	—	—	45	14	20	10	—	—	10	14	—	—	—	—	—	—	10	6	—	—	258	14
8.	Tardeo-Byculla	130	70	110	-	30	190	45	75	—	—	—	25	—	—	15	75	30	40	16	—	—	95	—	20	25	—	1091	16
9.	Worli-Elphin-Stone Chinchpokli	15	40	30	16	—	20	—	—	30	—	15	—	—	—	—	30	—	—	—	—	—	—	—	—	—	—	196	8

(Contd.)

TABLE 6.2 : *(Contd.)*

	1	2	3	4	5	6	7	8	9	10	11	12	13	14	15	16	17	18	19	20	21	22	23	24	25	26		
10. Dadar-Mahim	140	145	215	150	90	10	125	35	80	37	51	20	30	30	—	167	—	15	35	—	10	—	—	9	—	—	1395	19
11. Parel-Lalbaug	35	45	75	45	6	15	35	25	25	20	27	6	15	10	—	—	20	4	—	—	—	—	—	—	—	—	408	16
12. Matunga Sion	14	5	40	5	5	—	—	8	22	7	5	—	13	—	—	—	—	—	—	—	—	—	—	—	—	—	124	10
13. Bandra, Khar, Santacruz	20	55	108	30	55	12	25	20	72	34	5	—	34	10	—	—	7	—	10	4	34	10	18	—	5	—	568	20
14. Bandra Colony (E)-Kalina	3	20	—	—	—	—	10	5	—	4	—	—	3	—	—	—	—	—	—	—	—	—	—	—	—	—	45	6
15. Andheri-Juhu-Vile Parle (W)	20	39	55	13	36	15	29	13	11	13	5	9	12	5	—	—	12	-	6	5	—	—	10	—	—	—	308	18
16. Vile Parle(E)-Andheri-Jogeshwari	10	20	35	—	11	50	15	10	10	12	9	10	13	9	8	40	8	—	—	3	6	8	5	—	—	—	300	21
17. Goregaon-Malad (W)	105	85	160	30	24	25	37	10	45	25	—	26	—	7	40	—	5	20	—	3	30	58	60	—	—	—	795	19
18. Aarey, Malad, Dindoshi	10	65	105	35	55	5	40	—	—	—	—	15	—	—	—	—	—	5	—	—	—	—	—	—	—	—	335	9
19. Chembur-Kurla	85	85	144	25	33	95	80	30	27	18	6	4	40	10	23	46	—	18	—	8	—	20	10	—	—	—	807	20
20. Ghatkopar-Vikhroli	54	24	52	20	—	—	18	10	—	—	—	5	40	2	9	10	10	—	3	17	—	47	—	—	—	—	321	15
21. Borivli(W) Manori etc.	12	14	34	24	23	—	15	—	—	9	—	—	39	—	10	28	13	4	8	—	5	—	—	—	—	—	238	14
22. Poisar, Akurli, Dahisar	4	—	—	5	—	—	—	6	—	—	—	—	4	—	—	—	—	—	—	—	—	—	—	—	—	—	19	4

Source: Data collected from field investigation (1985).

Hierarchy level II with 14 to 19 commodity functions is associated with:

Dadar (W)-Mahim.
Goregaon-Malad (W)
Juhu-Anheri (W)
Colaba,
Tardeo-Byculla
Parel-Lalbaug,
Ghatkopar-Vikhroli
Borivli-Manori (W) and
Malabar-Cumballa Hill,
(In all 9 zones)

Hierarchy level III with 7 to 13 commodity functions is associated with:

Mandvi-Chakala,
Girgaum-Chowpatty,
Matunga-Sion
Aarey-Dindoshi-Goregaon-Malad (E), and
Worli-Chinchpokli.
(In all 5 zones.)

The remaining three zones, namely, the —

North and South Dock,
Bandra (E)-Kalina, and
Poisar-Akurli-Dahisar (E)

have the least number of commodity functions—less than 7. They belong to the lowest hierarchy level IV.

It may, however, be pointed out that the lesser number of functions in any retail zone does not indicate a higher degree of specialisation. In fact, the number of commodity functions seems to indicate the relative level of development and maturity of retail area. Hence, the lesser number of commodity functions gives an impression of an infantile stage of retail clusters.

It may also be noted that the higher number of commodity functions in hierarchy levels I and II is mainly associated with proximity to residential zones barring the case of Fort. But the converse is not true that the lower the commodity functions the farther it is away from the residential area. It also appears that the quality of the residential area determines the quality and level of commodity functions.

6.2 Quantitative Scores of Hierarchic Ordering of Retail Cluster Zones

6.2.1 *The Weighted Score Method*

The above stated qualitative analysis does not give any indication about the relative degree of extent of development of retail functions in any retail cluster. In order to evaluate the relative importance of the individual retail commodity functions and of the retail area, a weightage score was evaluated by allotting a weightage score to each retail commodity group by using the following formula:

$$\frac{N}{F_i}$$

where N is total number of zones and F_i is the number of zones having the specific function (i.e., commodity group). Weightage was given to various commodity groups (i.e., functions) based on the concept of "Greater the scarcity, the greater is the importance in terms of centrality and hence it got a higher score."[6] If a particular zone has x number of functional units of the specific commodity group F_i, weightage score for that function in that zone will be:

$$\Sigma \left(\frac{N}{F_i} \times x \right)$$

Summing up the individual scores, the composite functional score for each zone has been derived as shown in Table 6.3. Finally, the functional scores have been ranked in a descending order. Based on this score, five-level hierarchy has been obtained for Bombay and depicted in Fig. 6.1

The five-level retail hierarchy that has emerged distinctly reveals the primacy of the highest order obtaining in Bhuleshwar-Market and Fort through a very high weighted score and a clear break with the next level by a wide margin of score value. The next level is the wholesale commercial core at Mandvi-Chakala, and predominantly residential, old cores of Dadar-Mahim, Tardeo-Byculla, Girgaum-Chowpatty and a relatively recent development in the suburban zone of Goregaon-Malad (W). The occurrence of Goregaon-Malad in this group is apparently inexplicable but in all probability is due to the fact that the area is predominantly populated by the merchant community of Gujarati origin who have set up local shops as well. This hierarchy level grades into the next level through a imperceptable break not so well-defined as the break

TABLE 6.3: Composite Table on Retail Hierarchy in Bombay

Name of Zone	Zone No.	Retail Functional Index	Ranking Acc. to Retail Functional Index	Weighted Score	Ranking Acc. to Weighted Score	Index of Retail Strength	Ranking Acc. to Retail Strength	Over all Retail Hieratchy
Bhuleshwar	5	442.26	1	4462.75	1	130.85	2	
Fort	2	427.99	2	4092.35	2	133.65	1	I
Mandvi-Chakala	3	227.95	3	2578.12	3	99.53	4	
Tardeo-Byculla	6	219.68	4	1927.74	5	68.18	7	
Dadar-Mahim	10	204.13	5	2015.07	4	73.42	5	II
Goregaon Malad (W)	17	152.70	6	13.03.71	7	41.84	9	
Girgaum-Chowpatty	8	139.93	7	1771.03	6	100.46	3	
Colaba	1	112.86	8	1073.59	9	45.61	8	
Chembur Kurla	19	110.72	9	1163.14	8	40.35	10	
Bandra-Santacruz (W)	13	107.89	10	912.53	10	25.40	12	III
North South Dock	4	69.89	11	792.15	11	73.33	6	
Ghatkopar-Vikhroli	20	60.08	12	498.71	13	21.40	15	
Parel Naigaon	11	57.50	13	566.81	12	25.50	13	
Vile Parle-Andheri (W)	15	57.17	14	438.84	16	17.11	17	
Vile Parle-Andheri (E)	16	49.17	15	457.73	14	14.29	19	IV
Borivli (W)	21	44.19	16	363.44	19	17.00	18	
Malabar-Cumbala Hill	7	38.20	17	390.07	18	18.43	16	
Worli-Elphinstone Chinchpokli	9	29.46	18	456.12	15	24.50	14	
Goregaon-Malad (E)	18	27.66	19	417.50	17	37.22	11	
Dadar (E) Sion	12	19.61	20	166.77	20	13.66	20	
Bandra Santacruz (E)	14	6.20	21	58.73	21	7.40	21	V
Poisar-Akurli-Dahisar (E)	22	3.97	22	25.62	22	4.71	22	

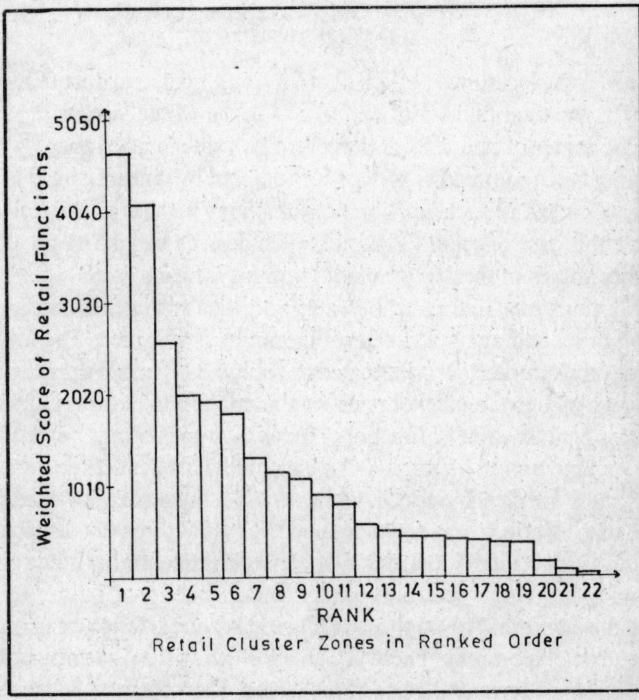

Fig 6.1: Hierarchic Grouping of Retail Zones based on Weighted Score Values of Retail Function

between the first and second levels.

The next two levels of retail hierarchy III and IV show a still more gradual and subtle variation in score values, the breaks between II and III and III and IV levels being less distinct. However, the lowest (V) level is marked by a relatively distinct break from level IV.

6.2.2 *The Centrality Index*

Yet another quantitative method of establishing retail hierarchy is based on centrality index.

The centrality index technique developed by Davies (1967)[7] has been used here for deriving the functional indices of different shopping centres and their hierarchic levels. Based on the number of retail establishments, under each retail type, the centrality index of a single establishment, called a location co-efficient, is computed by using the following formula:

$$C = \frac{t}{T} \times 100$$

"where C is the location co-efficient of funtion t, t is one outlet of function t, and T is the total number of outlets of t in the whole system".[8]

The degree of centrality or the centrality value for each zone for each functional type (commodity group) was derived by multiplying C by the number of outlets, of each functional type existing in the area. A functional index in this case of retail function is then derived by adding up all the centrality values of the 26 commodity groups for each zone.

The functional indices of all 22 zones are then arranged in descending ranking order and again a five level hierarchy is obtained. The level of hierarchy thus worked out is indicated in Table 6.3. A scatter diagram (Fig 6.2) drawn with the number of retail establishments in clusters plotted on the x-axis against value of functional index on y-axis brings out the five-level retail functional hierarchy. Once again the breaks between I and II and IV and V are clearly defined, while the other breaks are more gradual.

A map (Map 6.1) was drawn to show the spatial pattern of distribution of the intra-urban retail hierarchy on the basis of the centrality index value. It shows a very high centrality index (retail functional index) in and around the city core. This high centrality index zone is followed by a zone of second level hierarchy. There is a sharp break in the centrality indices in the western and central parts of the island. The centrality index value is comparatively higher in the eastern part of the city, the reason being that the eastern part of the island is the seat of wholesaling and warehousing activities. Retailing has developed in the eastern part as an associated function of wholesaling. In the northern part of the island around Dadar, a high retailing index zone is seen. The area has developed at the node of an intra-urban transport focal point.

In the suburbs, the highest retail index is seen in Goregaon-Malad (W) area. The next level is seen in Bandra-Santacruz (W) and Chembur-Kurla areas. In all other area, the retail hierarchy is low. The lowest hierarchy is seen in Bandra-Santacruz (E) Kalina, and Borivli-Poisar-Akurli-Dahisar (E).

6.2.3 *A Comparison of the Results obtained through Application of the two Techniques*

The same five-level retail hierarchy emerges from the application of the method of centrality index as with the weightage score method. The retail

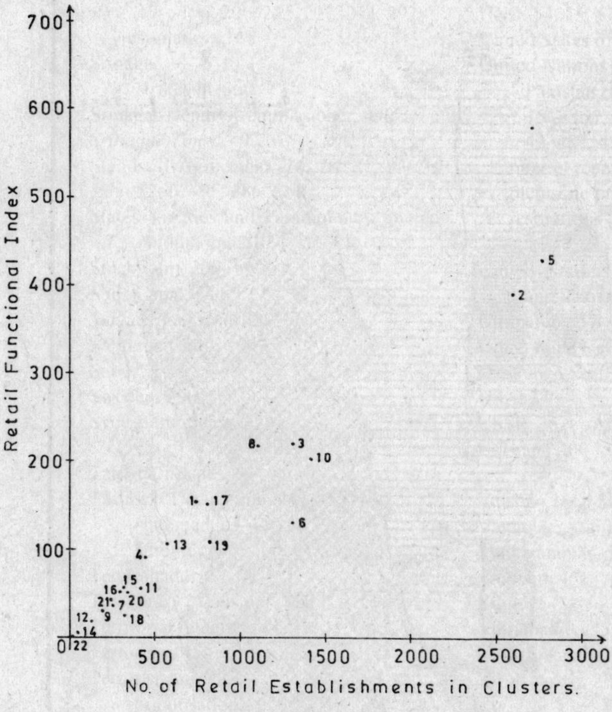

Fig 6.2: Scattergram of Retail Functional Index against No. of Retail Establishments in Cluster

cluster zones that fall within the hierarchy level by both methods are the same area, although within an individual hierarchic level, there are marginal shifts in position of zones in the ranking order. It can be inferred from the scattergram as well as from the map that the range of variation of centrality index between the highest and lowest zone is very high. The highest functional index of 442.26 for Bhuleshwar is nearly 111 times as high as the lowest score of 3, obtained for Borivli (E).

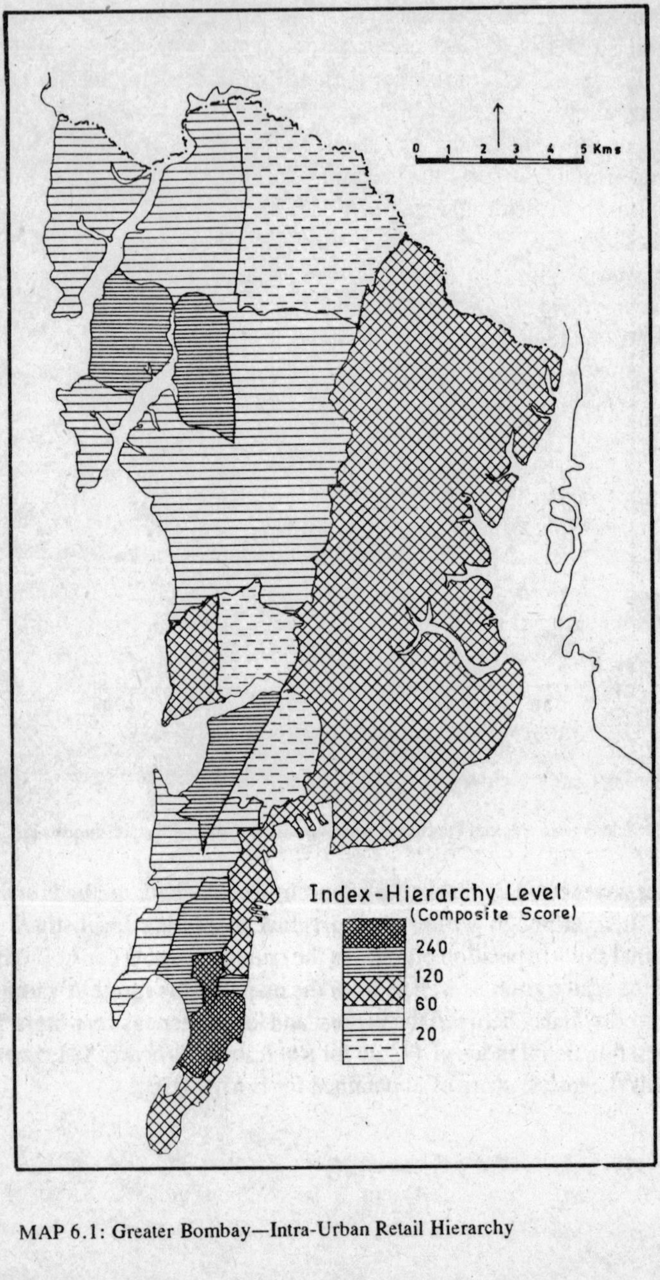

MAP 6.1: Greater Bombay—Intra-Urban Retail Hierarchy

6.2.4 *Retail Strength Method*

Yet another method of establishing retail functional hierarchy is based on 'Retail Strength'.

Garner's[9] study of the hierarchy of retail nucleation in Chicago has already been discussed. Garner's hierarchy of retail nucleation was determined by the analysis of the relationship between the number of occupied establishments and the number of business types. An attempt has been made to determine retail hierarchy of Bombay using Garner's method.

Accordingly, the number of retail establishments in a cluster zone has been divided by the types of retail functions (i.e., commodity groups) in that zone to derive an index of retail strength. These indices of retail strength have been arranged in descending order and once again a five level retail hierarchy has been obtained as shown in Table 6.3.

Based on Garner's method of determining retail functional index, a five-level retail hierarchy emerges for Bombay, albeit with little more variations in between the zones for certain zones compared with the first two methos discussed above, e.g., North and South Dock which received 11th rank by the first two methods but 6th rank by Garner's method.

A scatter diagram (Fig. 6.3) with the number of retail establishments in clusters plotted on the x-axis against value of retail strength on y-axis brings out a five-level retail functional hierarchy.

Under this method also, Fort scoring a retail strength index of 133.65 ranks first and Bhuleshwar with 130.85 ranks second; they constitute the retail hierarchy level I. In other hierarchic levels also, there are intrazonal shifts in ranking position of individual cluster zones, but the overall hierarchic levels remain practically the same.

Having worked out the retail hierarchy level by three different methods, the ranking order obtained by employing the three methods is shown in Table 6.3. It clearly reveals that irrespective of the method applied for determining the hierarchy, the hierarchic levels of retail cluster zones remain the same although there is a marginal shift of hierarchy within the groups.

6.3.1 *Association among Retail Parameters*

Having established a five-tier retail functional hierarchy in Greater Bombay, it is necessary to examine the relationship between this retail hierarchy and different urban and commercial parameters. An attempt is, therefore, made in this section to study the various facets of retail functions

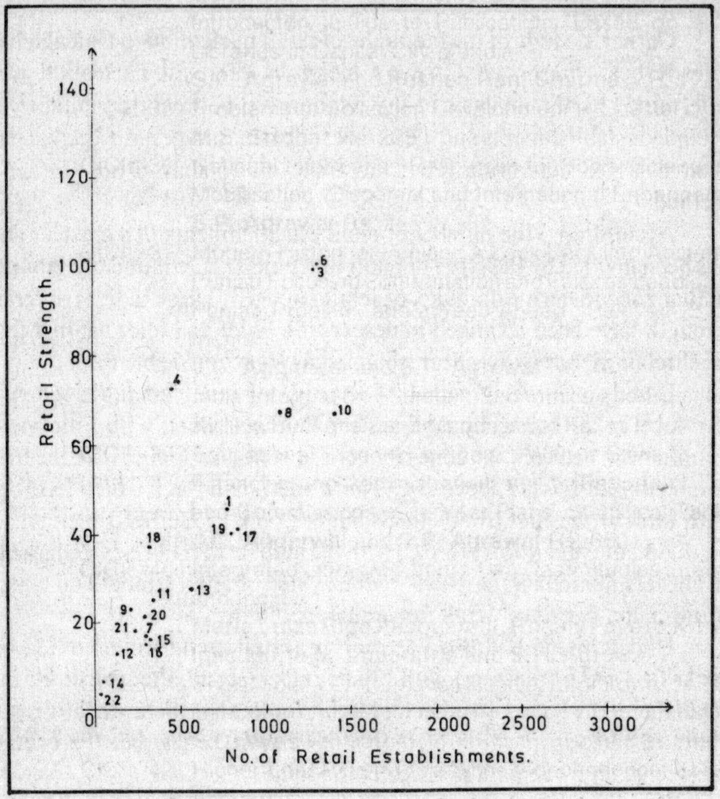

Fig 6.3: Scattergram of Retail Strength against No. of Retail Establishments

in the urban system of Bombay before moving over to the in-depth case studies.

Accordingly, it is attempted to study correlations between (i) retail establishments and retail functional index; (ii) total floor area and retail functional index; (iii) retail floor area and retail establishments; and (iv) population density and retail functional hierarchy.

6.3.2 *Correlation between Retail Establishments and Values of Retail Functional Indexes*

The relationship between number of establishments and retail index values as calculated is plotted on the scattergram shown in Fig. 6.4. It is a plot of the number of retail establishments in clusters against the corresponding retail functional index values. The scatter broadly reveals a linear relation. The scatter of log of the number of retail establishments in cluster in the 22 zones plotted on x-axis against log of retail functional index for 22 zones on y-axis also shows an almost perfect linear relationship (Fig 6.4)

The regression line of y on x has also been obtained in the usual form of the relation[10]

$y = a + bx$ (1)

By solving the normal equation (1) simultaneously the values obtained are:

$$a = -0.71 \text{ and } b = 0.968$$

The regression line equation is thus given by:

$$y = a + bx \text{ or } (-0.71) + 0.968\, x,$$

or better stated as

$$y = 0.968\, x - 0.71$$

It is seen from the regression equation and the scatter that there is a distinct positive linear relationship between the number of establishments and the retail functional index, i.e., the retail functional index value increases in proportion to the number of retail establishments in the clusters, or in other words, the retail functions performed in any cluster increase in complexity with an increase in the number of retail establishments occurring in the cluster.

Estimation of Standard Error (SE)

If Yest represents the value of y for given values of x, a measure of the scatter about the regression line of y on x is supplied by the quantity.

$$Sy, x = \sqrt{\frac{\Sigma (y\text{-Yest})^2}{N}} \qquad (2)$$

which is called the standard error of estimate of y on x.

By substituting the values in (2) we get

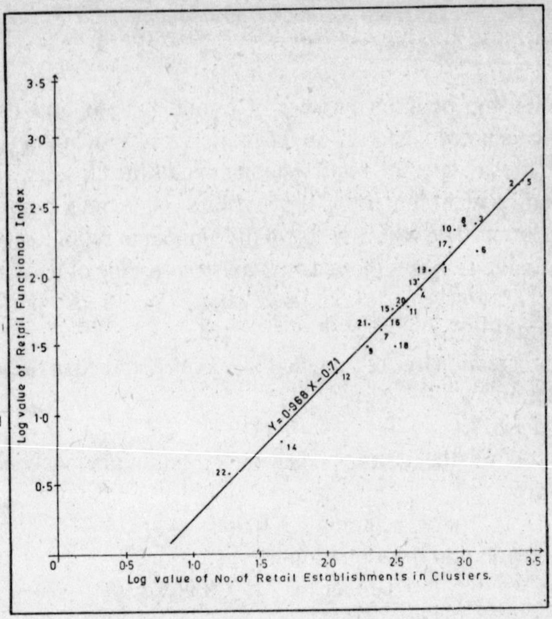

Fig 6.4: Regression between the Retail Functional Index and No. of Retail Establishments in Clusters.

Sy, x = 0.0

Thus, the standard error of estimates of *y* on *x* is zero.

The co-efficient of correlation *'r'* is given by the formula

$$r = + \frac{\text{explained variation}}{\text{Total of the (unexplained + explained variation}}$$

$$= + \sqrt{\frac{\Sigma\,(\text{Yest} - y)^2}{\Sigma\,(y - y)^2}}$$

By substituting the values in (3) we get explained variation Σ (Yest - $\bar{y}$ to be 5.41, while unexplained variation (residuals) $\Sigma\,(\text{Y} - \text{Yest})^2$ is 0.0 Therefore, by substituting the values in (3) we get

$$r = + \frac{5.41}{5.41} \text{ or } +1$$

This would mean that the two variables show a perfect positive correlation.

6.3.3 Correlation between Total Floor Area and Retail Functional Index

It was also attempted to correlate the total floor area (in thousand sq ft) of the zones of the study area (city only) by plotting on the x-axis in the scatter diagram against retail functional indices of these zones on y-axis (Fig. 6.5).

Since the floor area of the individual retail zones in the suburbs is not available, this relationship study is confined only to the city area, namely, the 12 retail cluster zones covering the city.

The equation of regression line is given by

$$y = 242.22 - 0.021 \quad (x)$$

The co-efficient of correlation between the total floor area and the retail functional index is found to be + 0.18. This weak association is obviously due to the fact that the total floor area is under varied types of urban land-use and not necessarily the retail floor area.

The regression residuals (standard error of the estimate 137.35 in this case) of the retail functional index of the total floor area have been mapped as shown in Map 6.2.

It is seen from the map as well as from the graph that except for three zones, namely, Bhuleshwar, Fort (both on the positive side) and Matunga-Sion (on the negative side), all the zones lie within one standard error distance on either side of the regression line. Both Bhuleshwar and Fort have got high positive anomalies, clearly demonstrating that both these areas have for the total floor area an excessively high retail functional index, or in other words, excess occupation by the retail function, which is explained by the location of Bhuleshwar in the commercial core of Bombay and of Fort in the commuting head of the C.B.D. adjoining the rail-cum-bus terminals. What is perhaps not so readily explained is the marked (but not so marked as Bhuleshwar-Fort) negative residual of Matunga-Sion, a middle income residential zone that had the status of a suburb in the 1940s, though now converted into a city residential area. The nearness of Dadar retail market and development characteristic of a suburban order in earlier days, though now changing, probably explain the situation.

Ronald Jones[11] in a similar study has attempted to study Edinburgh on the basis of select types of shops and total floor area to establish retail hierarchy. However, in the present study of Bombay city (excluding suburbs), the retail functional index covering all shops but occurring in clusters has been used against total floor area, to get an idea about retail-

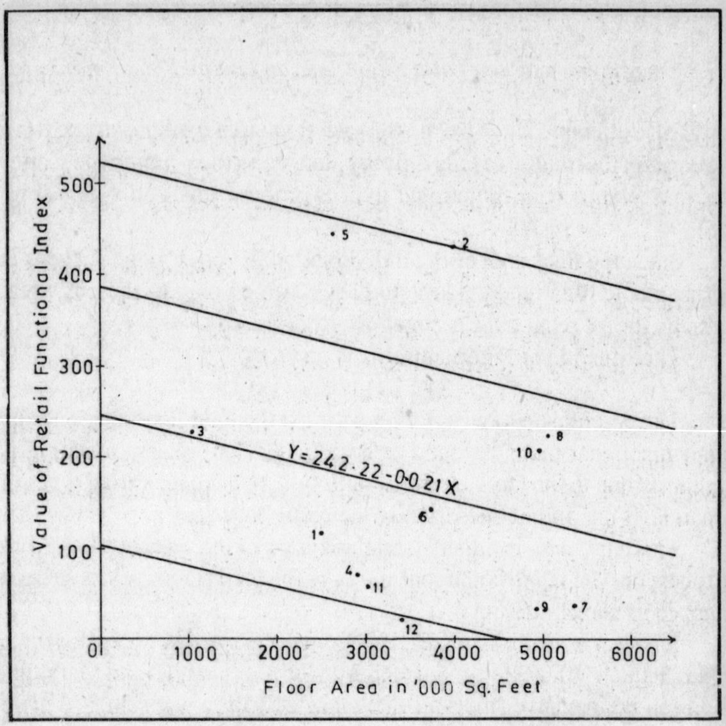

Fig 6.5: Regressin between Total Floor Area and Retail Functional Index

rich and retail-poor areas. Incidentally, the study shows that the city proper falls into four hierarchy zones of retail functions from the highest order, as follows.

I	High Positive (+1 to + 2 S.E.)	–	Bhuleshwar-Market, Fort.
II	Normal Positive (0 to + 1 S.E.)	–	Mandvi-Chakala, Tardeo-Byculla, Dadar-Mahim.
III	Normal Negative (1 to – 1 S.E.)	–	Colaba, North & South Dock, Girgaum Chowpatty, Malabar-Cumbala Hill, Worli-Chinchpokli, Parel-Lalbaug.
IV	High Negative (–1 to – 2 S.E.)	–	Matunga -Sion.

This hierarchy pattern is fairly consistent and comparable with the hierarchy pattern derived through employment of other techniques.

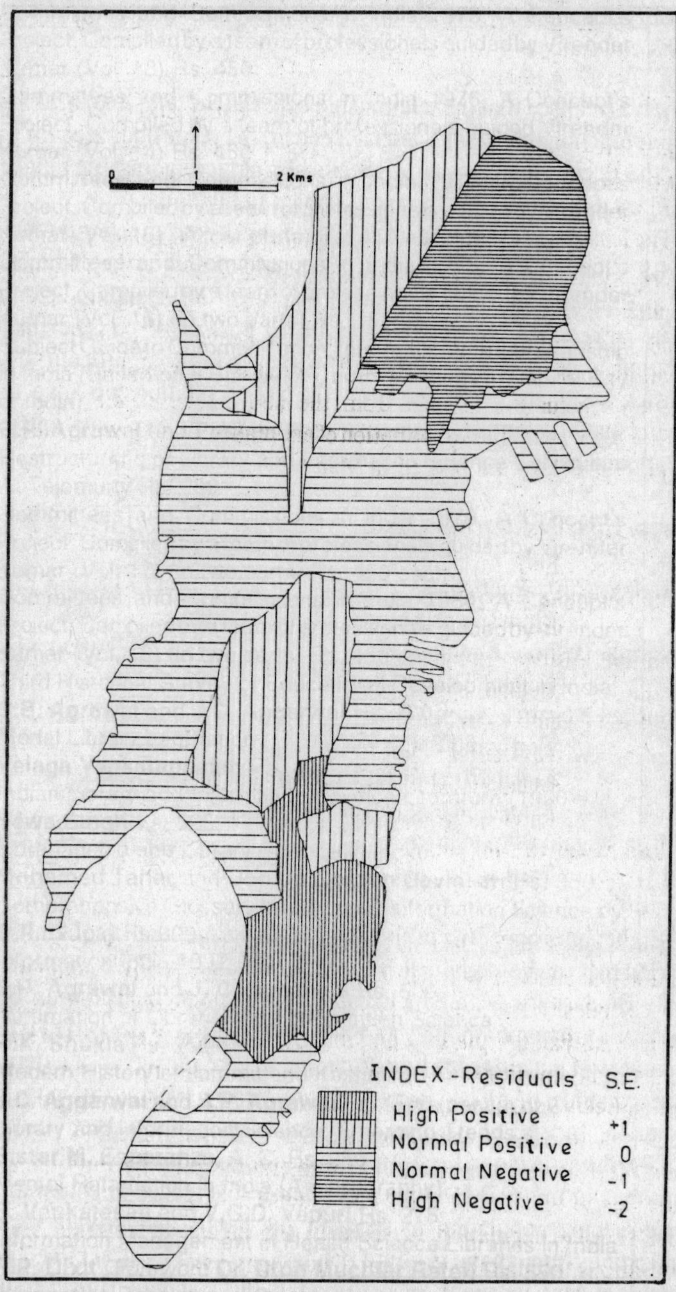

MAP 6.2: Bombay City: Regression Residual between Total Floor Area and Retail Functional Index

6.3.4 *Correlation between Retail Area and the Number of Retail Establishments*

Since it was realised that the relationship between the total floor area and the retail functional index is adequately clear, the retail floor area (in sq. ft.) for the 12 city zones was more specifically plotted against the number of retail establishments in clusters as shown in Fig. 6.6

The least square regression line is given by the equation $y = 0.00973 x - 20.91$, where y is the number of retail establishments in the 12 cluster zones and x is the corresponding retail floor area. The standard error of estimate *(syx)* works out to 671.82 and Karl Pearson's[12] co-efficient of correlation is of the value of +0.619, which is fairly significant. The regression residuals evaluated from the above relationship give the following spatial pattern of distribution of anomalies. Regression residuals, as mentioned below, are plotted in Map 6.3.

I Very High Positive Anomalies ________________ (> 2 S.E.):
Fort.

II High Positive Anomalies ________________ (+1 to +2 S.E.):
Bhuleshwar-Market, Mandvi-Chakala

III Normal Positive Anomalies ________________ (0 to +1):
Colaba, Dadar-Mahim.

IV Normal Negative Anomalies ________________ (0 to – 1 S.E.):
North and South Dock, Girgaum-Chowpatty,
Malabar-Cumbala Hill, Worli-Chinckhpokli
Parel-Lalbaug, Matunga-Sion.

V High Negative Anomalies ________________ (–1 to –2 S.E.)
Tardeo-Byculla.

Eight zones out of the twelve fall within normal statistical limits of deviation that can be explained. The areas of distinct positive anomalies, namely, Bhuleshwar-Mandvi-Chakala lie in the commercial land-use belt of the city's urban morphology and the concentration of the varied types of commercial functions in close proximity to take advantage of external urban economies in these zones seems to clearly explain this positive association. In case of the Fort area that reveals the highest positive anomaly, the association between administration, banking and financial institutions in this C.B.D. Zone and intense covergence of commuting service-seeking population to serve in the public and private sector organisations through the Western (Churchgate) and Central (V.T.) rail terminals and bus terminals seem to explain this situation. However, the distinct negative anomaly of the Tardeo-Byculla zone appears to be more difficult to explain. This is possibly due to lesser rental values in comparison to the city centre and its commercial use.

6.3.5 *Correlation between Resident Population Density and Retail Functional Index*

A correlation between resident population density and retail functional index was also attempted to examine the nature of relation between the two variables especially since retailing is a consumer service-oriented function.

The population density is plotted as the independent variable *(x)* and the retail functional index as the dependent variable *(y)*, shown in Fig. 6.7. Both are plotted in log values. Since the latter is available for all the zones of Greater Bombay, this plot covers the whole of the city and its suburbs.

The regression line of y on x has been obtained by using standard statistical relations, thus, in the form:

$$y = a + bx$$

$$y = 0.58 - 0.75$$

The standard error of estimate is obtained as +0.45 and Karl Pearson's co-efficient of correlation is +0.82, which is significant and high.

There exists a high positive linear correlationship between population density and retail functional index in 22 zones of the study area of Greater Bombay.

However, geographically, it is the areas of spatial deviations in the form of positive and negative residuals that are significant. The statistical distribution follows the normal distribution curve in that 72 per cent of the retail cluster zones lie within one standard error distance of the regression line. The regression residuals, which are plotted in Map 6.4 give the following spatial pattern of distribution of anomalies:

I Areas of Very High Positive Anomalies (>+2 S.E.)
Fort.

II Areas of High Positive Anomalies (+1 to +2 S.E.):
Goregaon -Malad (W).

III Areas of Normal Positive Anomalies (0 to +1 S.E.):
Colaba, Bhuleshwar-Market
Mandvi-Chakala, Dadar-Mahim,
Tardeo-Byculla, Ghatkopar-Vikhroli
Chembur-Kurla, Borivli (W), and
Juhu-Andheri-Versova.

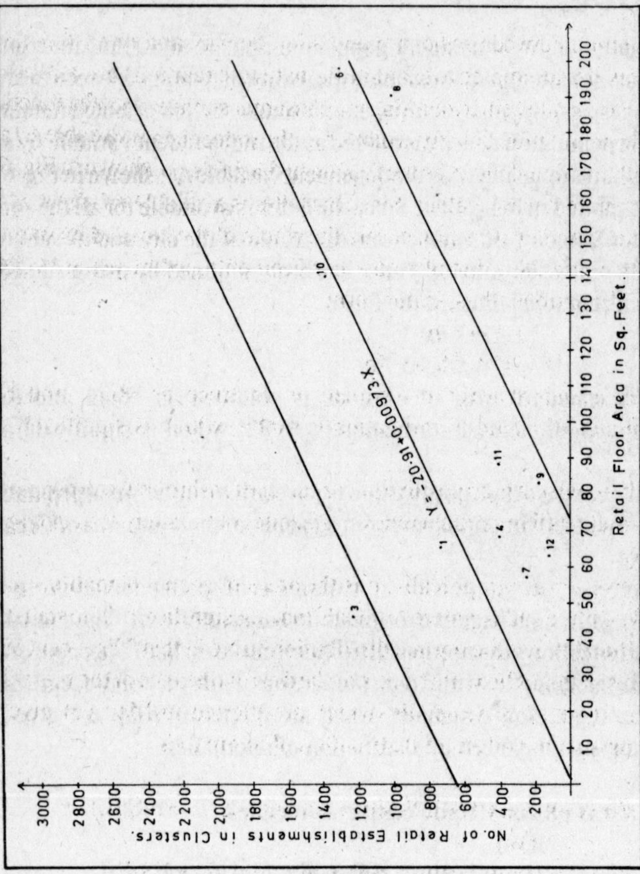

Fig 6.6: Regression between the Total Retail Area and No. of Retail Establishments in Clusters

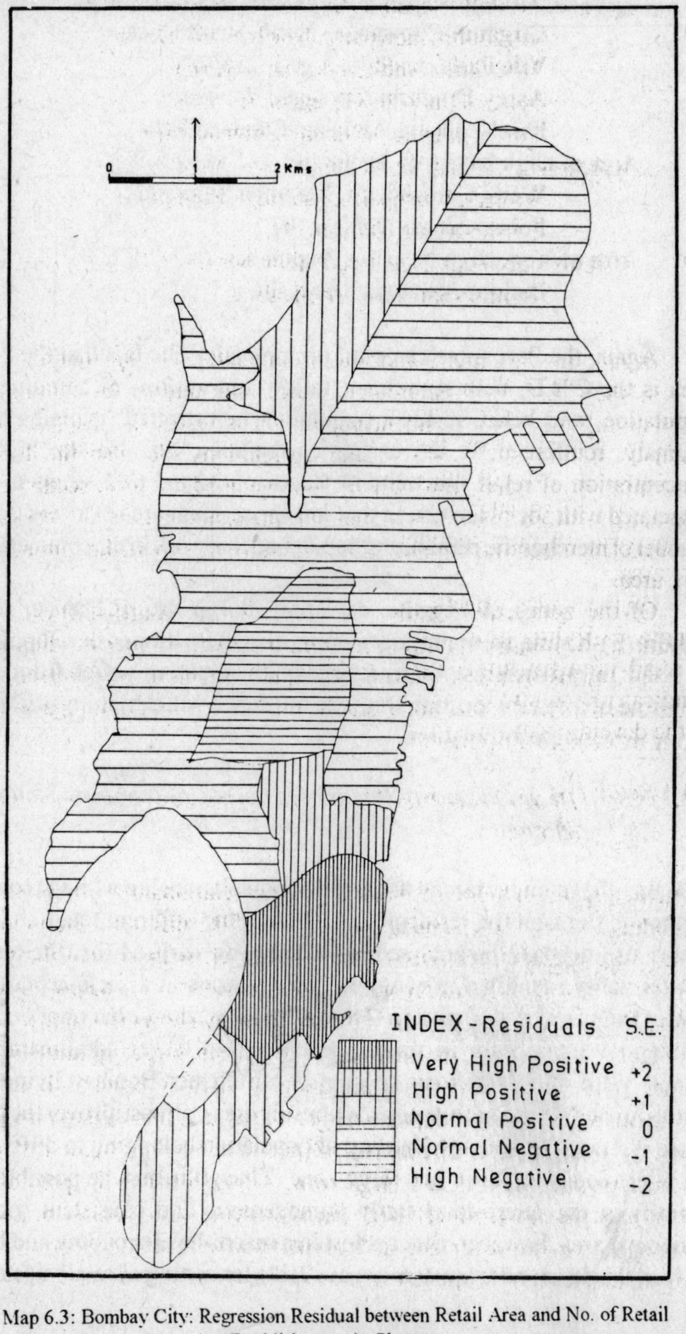

Map 6.3: Bombay City: Regression Residual between Retail Area and No. of Retail Establishments in Clusters.

IV Areas of Normal Negative Anomalies (0 to –1 S.E.)
Girgaum-Chowpatty, North-South Dock,
Vile Parle-Andheri-Jogeshwari (E)
Aarey-Dindoshi-Goregaon (E), and
Parel-Lalbaug, Malabar-Cumbala Hill.

V Area of High Negative Anomalies (–1 to –2 S.E.)
Worli-Chinchpokli, Matunga-Sion and
Poisar-Akurli-Dahisar (E).

VI Area of Very High Negative Anomalies (> – 2 S.E.):
Bandra-Santacruz (E)-Kalina.

Again, the Fort area stands out prominently. The fact that the Fort area is the C.B.D. with significant inflow and outflow of commuting population (and hence resident population is restricted) explains this anomaly, reinforced by the earlier correlations. So also the higher concentration of retail functions in Goregaon-Malad (W) seems to be associated with social factors, in that it is the resident area of a very large number of merchantile commuting population that work in the commercial core area.

Of the zones of negative deviation, Poisar-Akurli-Dahisar (E), Bandra (E)-Kalina are mainly developing areas with limited development of retail infrastructures. Worli-Chinchpokli seems to suffer from the handicap of a distant location from the rail-head, while Matunga-Sion is in the development transition.

6.4.1 *Retail Hierarchy and its Relations to the Socio-Economic Status of the Population*

Whether the retail hierarchy has any distinct relationship with the socio-economic status of the resident population of the different retail cluster zones also needs to be analysed. It is, however, realised that the retail cluster zones identified may not be homogeneous in a socio-economic sense. Empirical experience in Greater Bombay shows that upper class residential areas occur in small pockets within larger administrative zones. With almost half the population of Greater Bombay living in hutments and slums and dilapidated housing areas, almost all over the city there is a considerable intermixing of population belonging to different socio-economic strata in any large zone. Though it may be possible to identify at the micro-level fairly homogeneous and consistent socio-economic area, however, they get lost in a macro-level grouping and it is only at this level, relevant data are available for zoning of retail clusters.

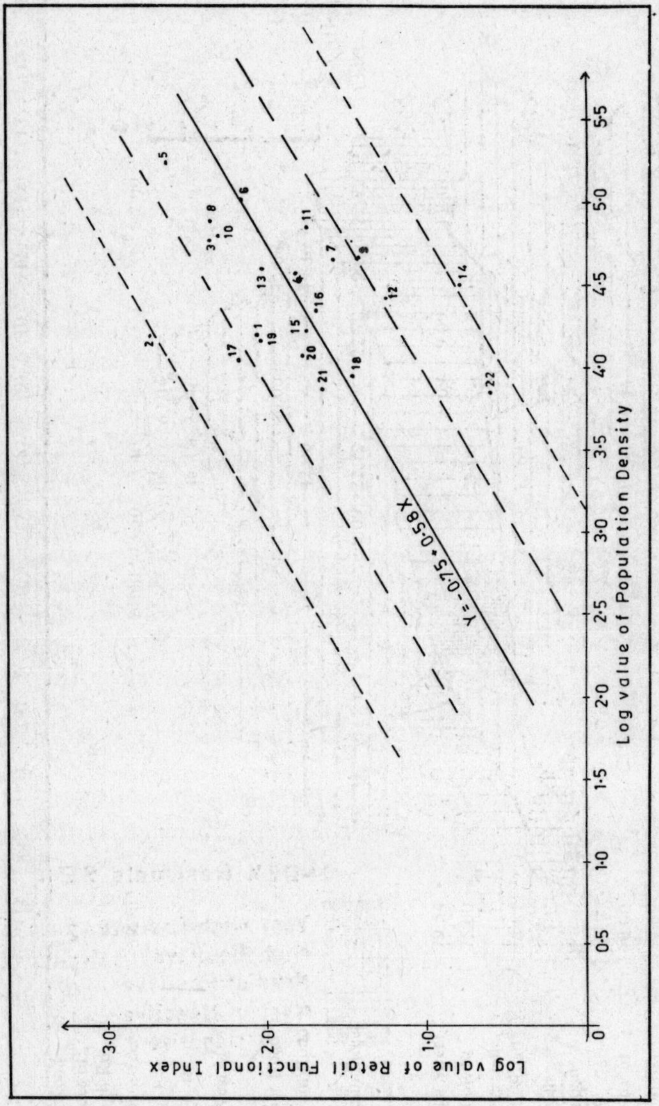

Fig 6.7: Regression between Population Density and Retail Functional Index

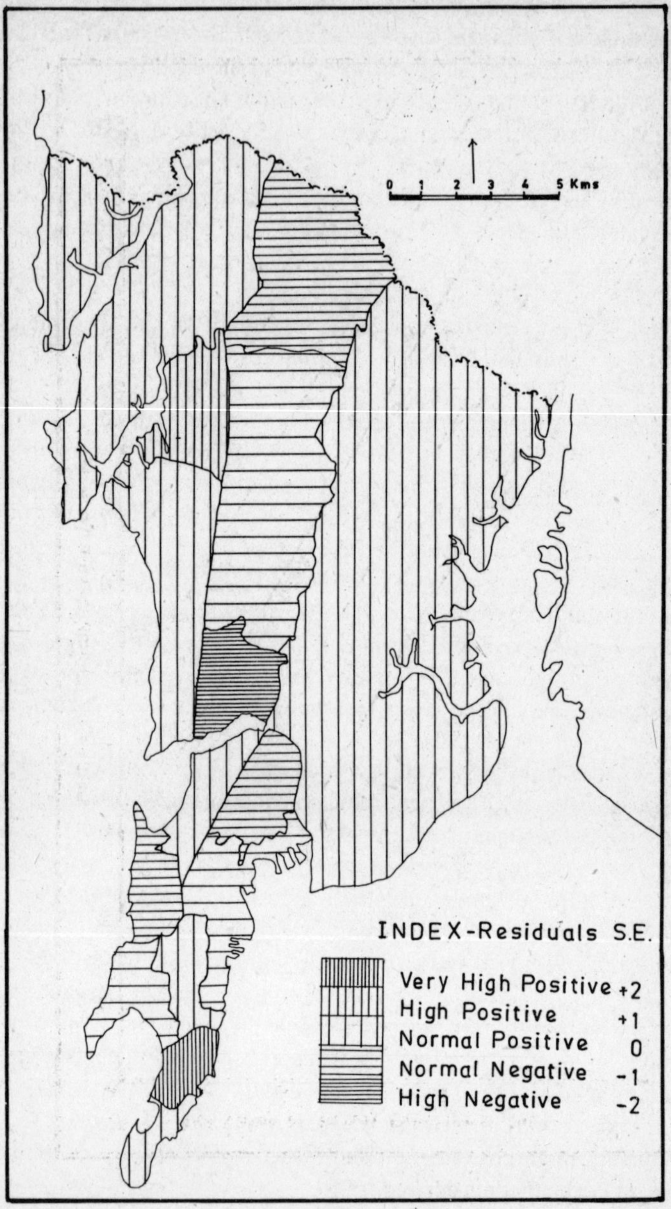

Map 6.4: Greater Bombay: Regression Residual Between the Population Density and Retail Functional Index

In fact, even the commercial and administrative zones of Bombay are not free from residence. By and large, all zones are socio-economically mixed, but it is pertinent to point out that the industrial areas are relatively more homogeneous and so too are some of the suburbs.

In view of this fact, to get a clear picture, it is proposed to work out a rank order of different retail cluster zones already identified on the basis of a number of parameters of socio-economic relevance.

For deriving a single composite standard score, 6 socio-economic indicators for each of the 22 zones of the study area have been used. They are:

1. Percentage number of households of the apartments and bungalows in the total number of households in the zones.
2. Percentage of households with telephones.
3. Percentage of households with automobiles.
4. Percentage of resident households living in ownership houses.
5. Percentage of resident households of huts in slum areas.
6. Percentage of households with a family income less than Rs. 1,000 per month.

It would be seen that the first four parameters are positive, since they have a direct positive association, while the last two indicators are negative in the sense that the relation between the indicator and socio-economic status is inverse. It may be pointed out that the choice of the indicators has been restricted on the basis of available data.

The data is drawn from a World Bank study conducted by Central Road Research Institute, New Delhi, prepared for the Bombay Metropolitan Region Development Authority[13] and Census of India, 1981.[14]

For deriving a composite score, standardisation of the data was effected through evaluating their percentage proportion to the total number of households as shown in Table 6.4. The percentage data for the indicators were ranked in a descending order as shown in Table 6.5.

Then the rank order for the four positive indicators for each zone was summed up; and so too the summation of the two negative indicators was done. The total score value of a particular zone was obtained by algebraic summing of the score values, positive and negative. On the basis of the composite rank values, the zones were ranked for establishing the socio-economic hierarchy. Table 6.6 shows the socio-economic hierarchy of the city of Bombay, thus derived.

It appears from this socio-economic hierarchy of Bombay that the Malabar-Cumbala Hill area has the highest rank followed by Bandra-

Khar-Santacruz (W), Fort, and Borivli(W).

In the next group fall

Girgaum-Chowpatty,
Juhu-Andheri-Versova, Bhuleshwar-Market,
Colaba and Goregaon-Malad (W), Matunga-Sion.

In the third zone are located

Tardeo-Byculla,
Poisar-Akurli-Dahisar (E),
Mandvi-Chakala, Dadar-Mahim,
Aarey-Dindoshi-Goregaon-Malad (E)

In the fourth zone are found

Ghatkopar-Vikhroli,
Bandra (E)-Kalina,
Chembur, Kurla,
Vile Parle-Andheri-Jogeshwari (E).

In the fifth and the lowest order are

Parel-Lalbaug,
Worli-Chinchpokli
North-South Docks.

The hierarchic pattern achieved is broadly consistent with the empirical experience. However, it appears to mark and underscore the impact of pockets of elite resident zones, as is seen in the case of Worli-Chinchpokli area. The socio-economic hierarchy is shown in Map 6.5.

The relationship between the socio-economic status (hierarchy) as derived above and the retail hierarchy as derived earlier was examined.

The scatter diagram (in Fig 6.8) plots the value of composite scores of socio-economic status of the 22 zones on the x-axis against values of retail functional indices.

The regression relationship between the two variables shown in the scatter is given by $y = 148 - 1.29x$. The standard error is 113, and the correlation co-efficient is +0.31, which is positive and weak.

From the regression residuals (Map 6.6) it is also noted that this relation brings out a five-level hierarchical zonation as follows:

I Very High Positive Anomalies ______________ (> + 2 S.E.):
Bhuleshwar-Market, Fort.

II High Positive Anomalies ______________ (+1 to +2 S.E.):
Mandvi-Chakala.

III Normal Positive Anomalies ______________ (0 to +1 S.E.):

TABLE 6.4: Percentage Distribution of Socio-Economic Indicators for Greater Bombay

Zone	% of Flats & Bungalows	% of Hutment Dwellers	% of House-holds with telephones	% of House-holds with Automo-biles	% of Owner ship Hou-seholds	% of Low Income House-holds
Colaba	36.99	31.19	31.19	23.04	25.19	67.59
Fort	60.50	1.79	25.46	28.02	18.79	58.89
Mandvi-Chakala	9.00	4.26	7.49	4.92	10.49	65.99
North & South Dock	22.99	17.00	2.50	2.54	5.09	82.49
Bhuleshwar Market	27.67	3.99	8.49	5.74	12.89	59.99
Girgaum-Chowpatty	24.99	0.43	9.49	6.93	13.49	54.99
Malabar-Cumbala Hill	69.99	2.99	42.59	43.77	33.59	36.69
Tardeo-Byculla	21.99	4.99	9.29	8.58	12.49	67.80
Worli-Chinchpokli	8.39	16.99	4.50	1.93	8.90	75.00
Dadar(W)-Mahim	26.29	20.99	3.09	3.42	13.29	64.29
Parel-Lalbaug	20.99	15.99	1.89	2.79	6.99	70.89
Matunga-Sion	37.99	21.99	18.30	7.97	12.86	62.20
Bandra-Santacruz (W)	63.99	0.70	16.19	15.36	49.89	50.60
Bandra Colony-Kalina (E)	.25	17.99	2.19	2.43	20.20	75.00
Juhu-Versova-Andheri (E)	42.99	16.99	3.99	8.16	33.69	60.00
Vile Parle-Andheri-Jogeshwar (E)	15.99	9.99	2.10	2.07	19.80	80.00
Goregaon-Bangurnagar-Malad (W)	41.65	10.80	2.29	4.48	35.79	65.00
Aarey-Dindoshi Goregaon (E)	23.99	7.99	2.19	4.05	26.59	73.19
Chembur-Kurla	22.99	23.99	1.09	4.25	21.35	73.00
Ghatkopar-Vikhroli	26.90	11.80	1.79	2.04	23.40	74.00
Borivli-Manori (W)	53.99	2.59	4.45	4.57	43.23	59.00
Borivli, Akurli Poisar, Dahisar	27.81	5.68	4.19	1.57	26.09	71.00

Source: Central Road Research Institute (1983) "Planning of Road System for Bombay Metropolitan Region Vol. I, Sec. IV)

TABLE 6.5: Derivation of Composite Score for 22 Zones of Greater Bombay

Sr. No.	Zones	% Ranking of Residents in Flats & Bungalows	% Ranking of Households with Automobiles	% Ranking of Households with Telephones	% Ranking of Ownership to Rental Households	Total	% Ranking of Hutment Dwellers	% Ranking of Households with Low Income	Total	Composite Score
(1)	(2)	(3)	(4)	(5)	(6)	(7)	(8)	(9)	(10)	(11)
1.	Colaba	8	3	5	8	24	1	12	13	11
2.	Fort	3	2	2	13	20	20	19	39	–19
3.	Mandvi-Chakala	21	10	9	19	59	16	13	29	30
4.	North & South Dock	16.5	17	15	22	70.5	6	1	7	62.5
5.	Bhuleshwar Market	10	9	8	16	43	17	17	34	9
6.	Girgaum-Chowpatty	14	8	6	14	42	22	20	42	0
7.	Malabar-Cumbala Hill	1	1	1	5	8	18	22	40	–32
8.	Tardeo-Byculla	18	5	7	18	48	15	11	26	22
9.	Worli-Chinchpokli	22	21	10	20	73	75	35	11	62
10.	Dadar-Mahim	12	15	10	15	52	4	10	14	38
11.	Parel-Lalbaug	19	16	20	21	76	9	9	18	58
12.	Matunga-Sion	7	7	3	17	34	3	15	18	16
13.	Bandra-Khar-Danda Santa-cruz (W)	2	4	4	1	11	21	21	42	–31
14.	Bandra Colony, Kalina(E)	13	18	17.5	11	59.5	5	3.5	8.5	51

(1)	(2)	(3)	(4)	(5)	(6)	(7)	(8)	(9)	(10)	(11)
15.	Andheri-Juhu-Versova (W)	5	6	13	4	28	7.5	16	23.5	4.5
16.	Vile Parle, Andheri, Jogeshwari	20	19	19	12	70	12	2	14	56
17.	Goregaon-Malad(W)	6	11	16	3	36	11	14	25	11
18.	Aarey-Dindoshi-Goregaon (W)	15	14	17.5	6	52.5	13	6	19	33.5
19.	Chembur-Kurla	16.5	13	22	10	61.5	2	7	9	52.5
20.	Ghatkopar-Vikhroli	11	20	21	9	61	10	5	15	46
21.	Borivli-Manori- (W) Gorai	4	12	11	2	29	19	18	3.7	–8
22.	Poisar-Akurli, Borivli (E), Dahisar	9	22	12	7	50	14	8	22	28

North-South Docks, Tardeo-Byculla, Dadar-Mahim, Goregaon-Malad (W), Chembur-Kurla.

IV Normal Negative Anomalies ________________ (0 to -1 S.E.)
Poisar-Akurli-Dahisar (E), Ghatkopar-Vikhroli, Aarey-Dindoshi-Goregaon-Malad (E), Juhu-Andheri-Versova (W), Vile Parle-Andheri-Jogeshwari (E), Bandra (E)-Kalina, Bandra-Khar-Santacruz (W), Matunga-Sion, Worli-Chinchpokli-Girgaum-Chowpatty Colaba, Parel-Lalbaug.

V High Negative Anomalies ________________ (-1 to -2 S.E.)
Borivli-Manori (W), Malabar-Cumbala Hill.

Fort and Bhuleshwar stand out prominently, being located at the city core with high intensity of business and commerce. Mandvi-Chakala is coming in the next level, because of high concentration of wholesale business activity.

Table 6.6: Socio-Economic Hierarchy of Greater Bombay

Zone No.	Rank	Name of the Zone	Composite Score Value	
7	1	Malabar-Cumbala Hill	–32	
13	2	Bandra-Khar-Santacruz (W)	–31	I
2	3	Fort	–19	
21	4	Borivli, Manori (W)	–8	
6	5	Girgaum-Chowpatty	0	
15	6	Juhu, Andheri-Versova (W)	4.5	
5	7	Bhuleshwar-Market	9	II
1	8.5	Colaba	11	
17	8.5	Goregaon, Malad (W)	11	
12	10	Matunga-Sion	16	
8	11	Tardeo-Byculla	22	
22	12	Poisar-Akurli-Dahisar (E)	28	III
3	13	Mandvi-Chakala	30	
18	14	Aarey-Dindoshi, Goregaon-Malad (E)	33.5	
10	15	Dadar-Mahim	38	
20	16	Ghatkopar-Vikhroli	46	
14	17	Bandra Colony-Kalina (E)	51	IV
19	18	Chembur-Kurla	52.5	
16	19	Vile Parle-Andheri-Jogeshwari (E)	56	
11	20	Parel-Lalbaug	58	
9	21	Worli-Chinchpokli	62	V
4	22	North and South Dock	62.5	

High negative anomalies are seen in Borivli (W) and Malabar Cumbala Hill. Both these areas are in the highest level of socio-economic hierarchy. Hence, retailing is not a predominat function in these areas.

6.5 Systematic Variation in Retail Functions within Greater Bombay—An Application of Distance Decay Concept

6.5.1 *Distance Decay of Population Density from City Centre*

It is an established truth that the distance factor from city centre plays an important role in explaining the spatial variations of urban functions. This

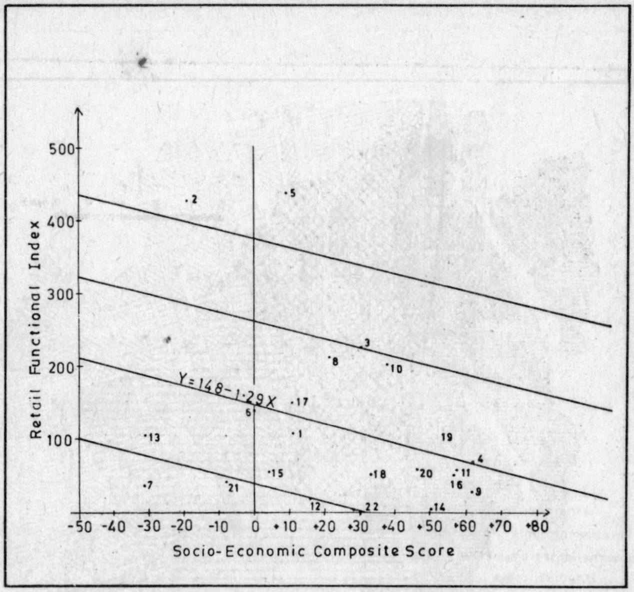

Fig. 6.8 : Regression between socio-economic Composite and Score and Retail Functional Index.

is often called the friction of distanc, interaction decay or the distance decay.

The distance decay from the city centre of Bombay and the spatial variation of population density as well as retail functional index of Bombay are now examined to test the mutual relationship between them.

It has been shown in earlier pages that the retail functional index is positively correlated with population density.

In Greater Bombay, the centre of density of population in the last eight decades has been steadily shifting from Central Bombay towards the north and in 1981 has moved into the southern part of suburban Bombay in Salsette.[15] This shows that more than 50 per cent of the population at present is in the suburbs. Yet, the population densities by and large decline from the city centre outwards. The most widely used density gradient model, the negative exponential model (Clark)[16], is applied to examine the distance decay of population density within Greater Bombay. The mathematical expression of the model is in the form.

$$y = \exp(a-bx)$$

where exp *(a)* is the density at distance x=0 from the city centre, or in other words the central density, and *b* is the constant indicating the rate of

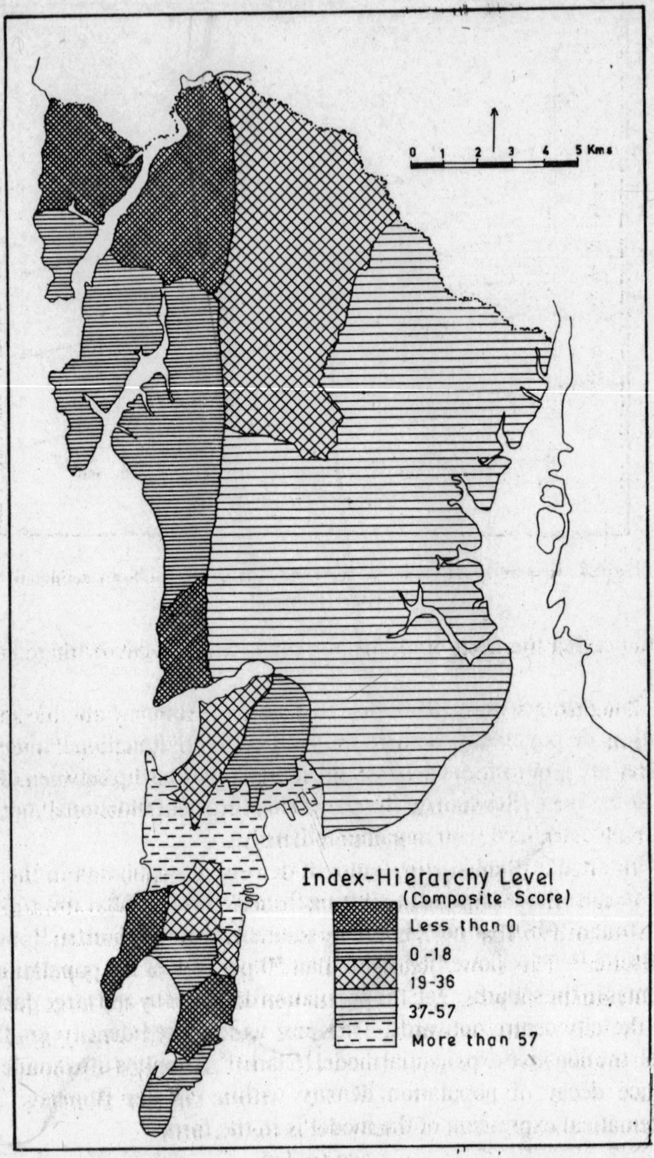

Map 6.5: Greater Bombay: Socio-Economic Hierarchy

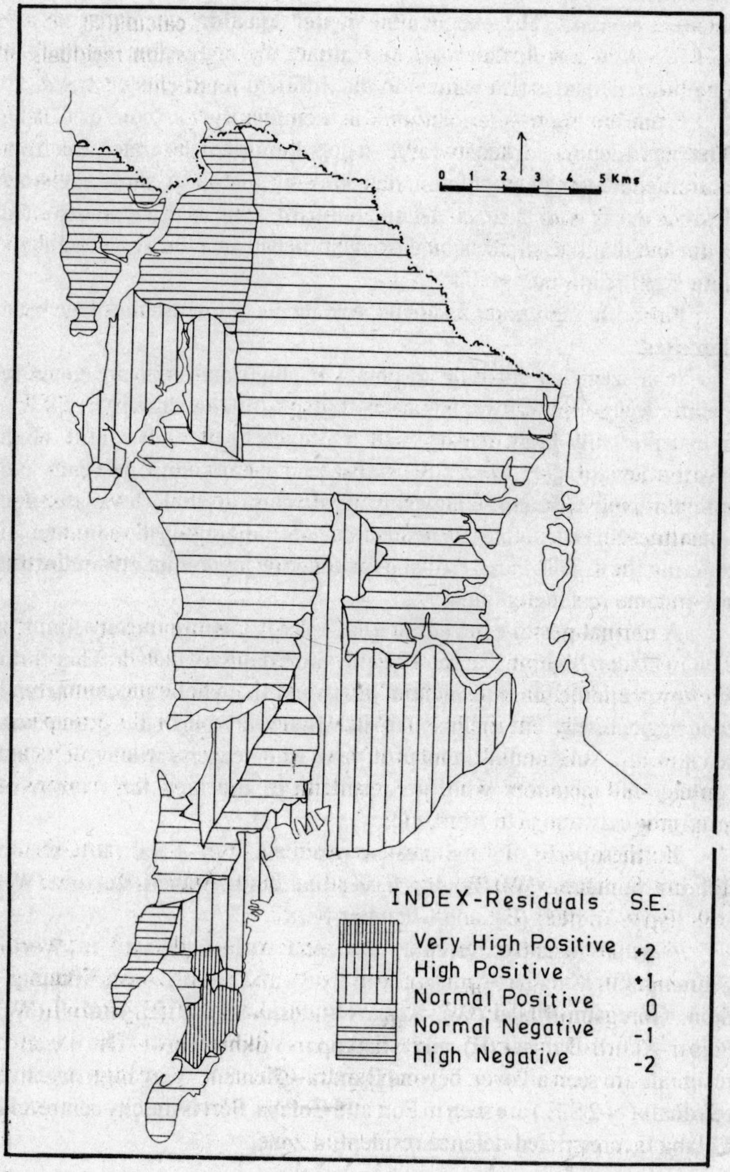

MAP 6.6: Greater Bombay: Regression Residual between Socio-Economic Score and Retail Functional Index

decline in natural logarithms. Fig 6.9 shows the scattergram wherein distance from the city centre (x-axis) is plotted against population densities (y-axis). The exponential model equation calculated as $y = 64215e - 0.06\,x$ is further used to evaluate the regression residuals of population densities (Ln values) in the different retail cluster zones.

From the regression relation it is seen that the city core or C.B.D. (Fort) has a depressed density value - a population density crater. But from its immediate fringe, population densities, by and large, show a distinct distance decay with a steady decline outward. This is also demonstrated by the fact that the co-efficient of correlation between the two variables is quite significant, i.e., +0.74.

From the regression residuals, five levels of hierarchies have been identified.

It is seen that spatially residuals of abnormal order are noted in specific zones. Bhuleshwar has a very high positive deviation (> +2 S.E.), or exceptionally high density with reference to the city centre. High positive deviation (+1 to +2 S.E.) is also seen in Girgaum-Chowpatty and Parel Lalbaug. Girgaum-Chowpatty is very close to Bhuleshwar area and constitutes the old, native city residential core, adjoinging the commercial zone and the C.B.D. Parel-Lalbaug includes the predominantly industrial low-income residential area.

A normal positive deviation (0 to +1 S.E.) within the city limits is seen in Dadar-Mahim, Tardeo-Byculla and Mandvi-Chakala. They form the lower middle class residential areas and the wholesale commercial zone respectively. The inclusion of the wholesale zone in this group is on account of a substantial number of petty shopkeepers, wholesalers and commercial labourers who are residents in the area for reasons of proximity-advantage to work site.

In the suburbs, normal positive residuals (0 to -1 S.E.) are seen in Bandra-Santacruz (W), Bandra (E)-Kalina, Juhu-Andheri-Versova (W), Vile Parle-Andheri (E), and Chembur-Kurla.

Normal negative residuals are seen within the city in Worli-Chinchpokli, Malabar-Cumbala Hill, North and South Docks, Matunga-Sion, Goregaon-Malad (W), Aarey-Dindoshi-Malad (E), Borivli (W), Poisar-Akurli-Dahisar (E), and Ghatkopar-Vikhroli areas. The negative residuals are seen all over, beyond Bandra-Chembur. Very high negative residuals (>–2 S.E.) are seen in Fort and Colaba. Fort is the city centre and Colaba is a restricted defence residential zone.

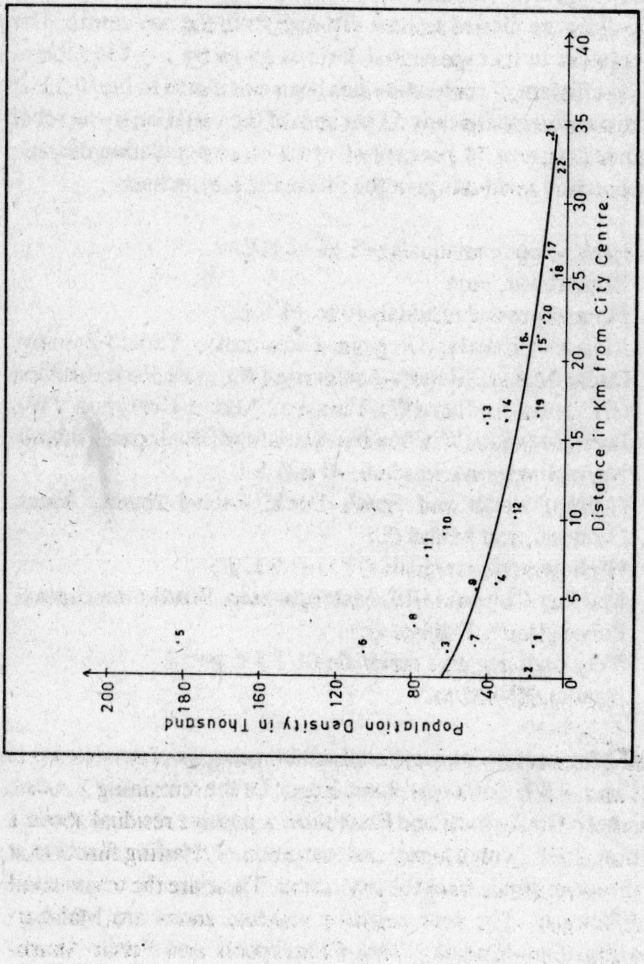

Fig 6.9: Scattergram of Population Density against Distance in km from City Centre

6.5.2 *Relationship between Distance from City Centre and Retail Functional Index*

A similar negative exponential model is applied to examine the relationship between the retail functional index and distance from the city centre on similar lines. Fig 6.10 shows the corresponding scatter when retail functional indices are plotted against distance from the city centre. The regression relation in its exponential form is given by y = 153.699e – 0.06x. The co-efficient of correlation has been calculated to be +0.53. It means that distance explains only 53 per cent of the variation in the retail functional index, against 74 per cent of variation in population density.

The regression residuals give the following hierarchies:

(i) High positive residuals (+1 to +2 S.E.):
Bhuleshwar, Fort.

(ii) Normal positve residuals (0 to +1 S.E.):
Mandvi-Chakala, Girgaum-Chowpatty, Tardeo-Byculla, Dadar-Mahim, Bandra-Santacruz (W), Vile Parle-Andheri (E), Juhu-Andheri(W)-Versova, Malad-Goregaon (W),. Borivli-Manori (W, Chembur-Kurla, and Ghatkopar-Vikhroli.

(iii) Normal Negative residuals (0 to 1 S.E.):
Colaba, North and South Dock, Parel-Lalbaug, Aarey, Dindoshi, and Malad (E)

(iv) High negative residuals (-1 to -2 S.E.):
Malabar-Cumbala Hill, Matunga-Sion, Worli-Chinchpokli, Poisar-Akurli, Dahisar (E).

(v) Very high negative residuals (>-2 S.E.):
Bandra (E)-Kalina.

15 out of the 22 zones lie within the explainable margins of deviation (i.e. between +1 and -1 S.E. from regression lines). Of the remaining 7 zones, 2 zones (namely, Bhuleshwar and Fort) show a positive residual above 1 S.E. but below 2 S.E. with a heavy concentration of retailing function at a 'stone's throw' distance from the city centre. These are the major retail clusters of Bombay. The four negative residual zones are Malabar-Cumbala Hills, Sion-Wadala, Worli-Chinchpokli and Poisar-Akurli-Dahisar (E). Very high negative residuals are seen in Bandra (E)-Kalina.

It is observed that the population density is very low at the city centre, followed by a peak at its periphery. Fort being the city centre has thus very high negative deviation and Buleshwar at the periphery, very high positive deviation. But the distance-decay function of retail index reaches

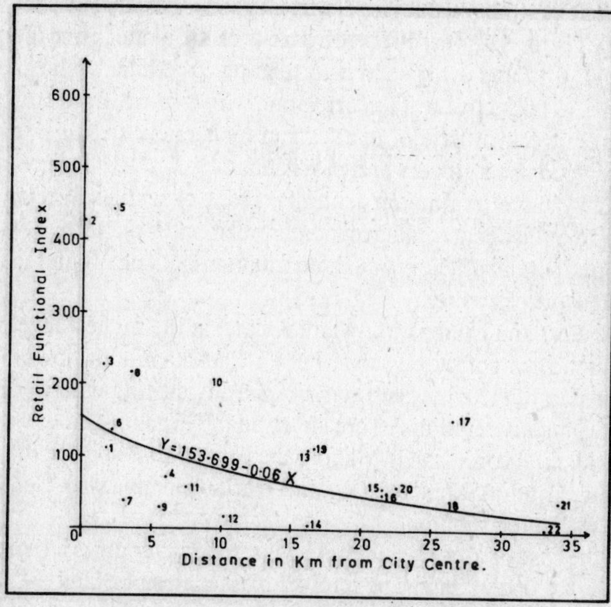

Fig 6.10: Scattergram of Retail Functional Index against Distance in km from City Centre

its peak at zero distance from the city centre and continues at peak level even in its periphery. Hence, Fort and Bhuleshwar have very high positive deviation. It is also noted here that at a distance of 9 kms both the residuals of density (0.52) as well as retail index (0.82) are positive. In case of population density after crossing the 10 kms distance the fall of the curve becomes gentler, but it is not so in the case of the retail functional index. Thus, within the overall distance density pattern, the density profile is uni-nodal with a creater at the centre. This crater is to a certain extent comparable with the western cities. This crater can be explained by the displacement of residential function by the non-residential activities. This crater in the city centre in case of city of Bombay simply implies population decline, but in no sense it is an economic decline.

The spatial variation of population density and retail functional index cannot thus be explained only by distance-decay. It has apparently a lot of other socio-economic implications.

6.5.3 The nature of relationship between distance from city centre and retailing function tested on the whole for retailing clusters was further tested with reference to three commodities, textile goods, grocery and provisions and electrical goods, that have a ubiquitous dispersal all over the city. The centrality values for electrical goods, grocery and provisions and textile goods as against distance from city centre plotted on scattergram (Figs. 6.11, 6.12 and 6.13) show a correlation co-efficient value of 0.51, 0.46 and 0.33 respectively. The correlation with distance from city centre is relatively weak in case of textiles and moderately strong in case of electrical goods and grocery and provisions.

From the scattergram on centrality index of textiles in relation to distance from city centre, an exponential relation in the form $y = 4.9e - 0.05x$ can be obtained. The residuals distribution statistically and spatially gives a four-level hierarchy (Fig. 6.13).

The Fort and Bhuleshwar-Market stand at the highest level being located at the heart of the city and Mandvi-Chakala is at the lowest order being the part of the wholesale market specially dealing with spices and grains. Secondary retail nodes are observed in Girgaum-Chowpatty, the older residential core, Dadar-Mahim, the commercial node in the north of the city, Bandra-Santacruz (W) and Kurla-Chembur and Goregaon-Malad (W), in the suburb.

In the case of grocery and provisions, the scatter gives a regression relation $y = 6.96e - 0.06x$. The semi-wholesale area of Mandvi-Chakala (Zone 3) with a large positive residual has a primacy and forms the highest level in the hierarchy. Fort and Market close to city core, Dadar-Mahim, Bandra-Santacruz (W) and Chembur-Kurla of the inner suburbs, Goregaon-Malad in outer suburbs, come out as distinct subsidiary retail nodes above the general distance-decay function (Fig. 6.12).

In the case of electrical goods too, the scatter reveals a general distance-decay function operating, but once again, already a prime node in the Fort and Market area is observed with secondary nodes at Dadar in the north of the city. Goregaon-Malad in outer suburbs stand out, with the lower retailing order at Girgaum-Chowpatty, Chembur-Kurla, Ghatkopar-Vikhroli, following suit, (Fig. 6.11).

The pattern obtained in the case of the three commodities that have a widely dispersed pattern of distribution is broadly consistent with the general pattern of distance-decay of retailing function as a whole in the city and the clear presence of primary and secondary retailing nodes catering to different parts of the city and its suburbs.

6.5.4 *From a careful scrutiny of the distance-decay of retail functional index, it may be summed up*

(i) Within the metropolis of Bombay, distance from city centre alone is not responsible for the decay of retail functional index.

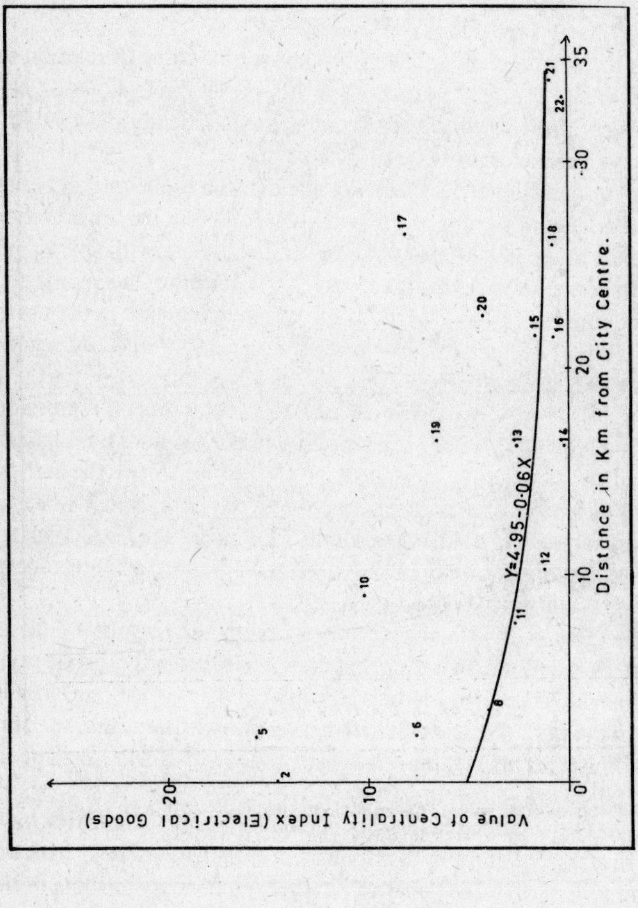

Fig 6.11: Scattergram of value of Centrality Index of Electrical Goods against Distance in km from City Centre

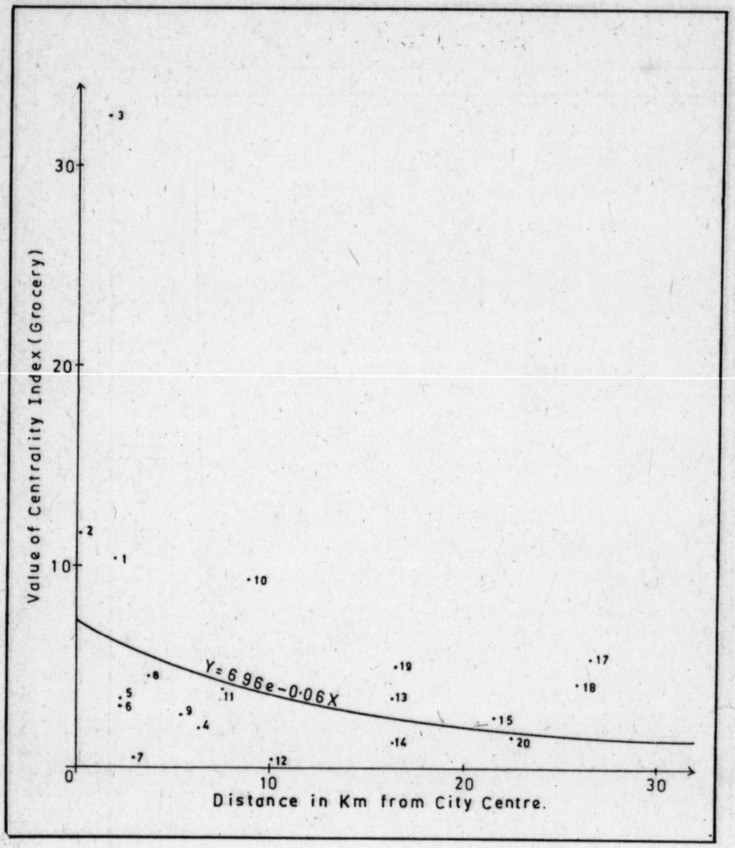

FIG 6.12: Scattergram of the value of Centrality Index of Grocery against Distance in km from City Centre

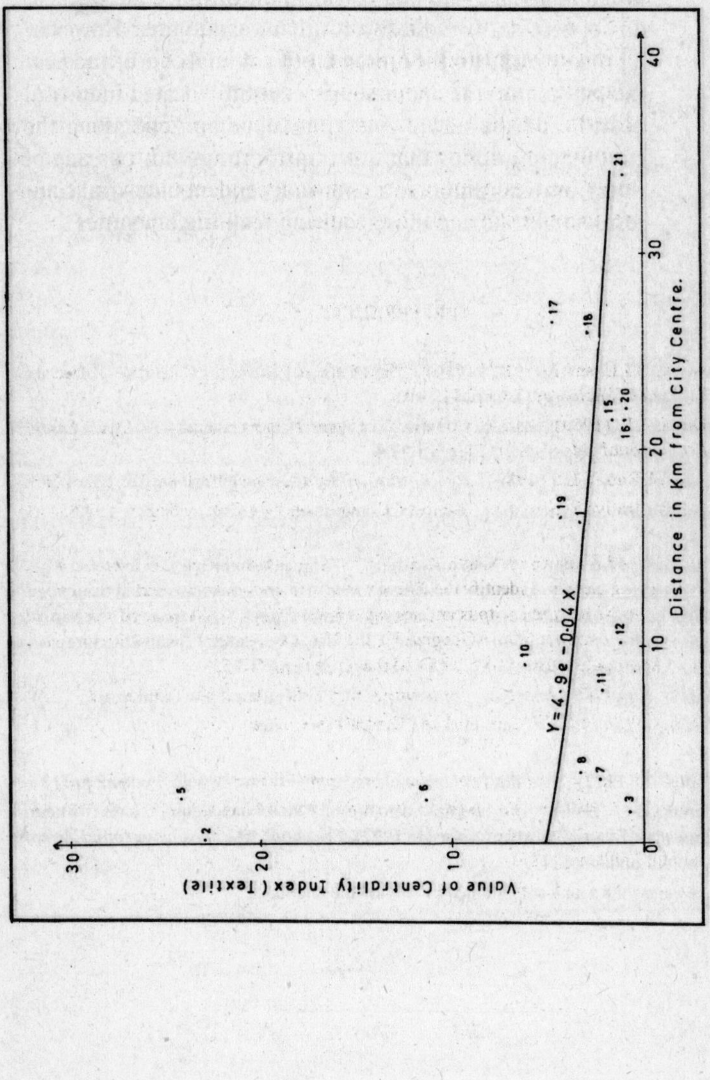

FIG 6.13: Scattergram of the value of Centrality Index of Textile against Distance in km from City Centre

The type of residential function and the history of development of retailing also influence retail functional index.

(ii) Though Greater Bombay–the city and its suburbs–spread to a distance of 34 kms from the city centre, it is a monolithic urban landscape; ups and downs in the distance-decay curve do not necessarily indicate any urban shadow area. However, it broadly confirms the presence of a colonial core at and near the city centre, the anomalously centrally located industrial hearth, and the newly emerging suburban zone along the commuting arteries that are transforming with passage of time from a dormitory to a semi-independent industrial cum-residential state and thus acquiring retailing amenities.

REFERENCES

1. Berry, B.J.L. and Rees, P.H. (1969): "The Factorial Ecology of Calcutta", *American Journal of Sociology*, 74, pp. 445-491.
2. Dutta, A.K. (1969): "Intra-city Hierarchy of Central Places, Calcutta — A Case Study", *Professional Geographer*, 21, pp. 18-23.
3. Prakasa Rao, V.L.S. (1983): *Urbanisation in India*, Concept Publishing Co., New Delhi.
4. Administrative Report of the Municipal Commissioner for Greater Bombay, 1980-81, p. 363.
5. This method is known as "Guttman Scaling", "The method which scales the retail type by shopping centres to identify the different levels of shopping centres and in the process points out the most ubiquitous and segregated retail types." — Extracted from Sastry, G.S., *Contribution to Indian Geography*, VII, Urban Geography, "Retail Structure of an India Metropolis—Bangalore", P.D. Mahadev (ed.) pp. 58-75.
6. Bhat, L.S. (1976): *Micro Level Planning*, K.B. Publications, New Delhi, p. 60.
7. Davies, W.K. (1967): "Centrality and Central Place Hierarchy", *Urban Studies*, pp. 61-79.
8. Carter, H. (1972): *The Study of Urban Geography*, Edward Arnold, London, p. 112.
9. Garner, B.J. (1966): "The Internal Structure of Retail Nucleation", North Western University, Evaston, cited by Carter, H. (1972): *The Study of Urban Geography*, Edward Arnold, London, p. 115.
10. The constant a and b are obtained by solving the relation

$$a = \bar{y} - b\bar{x}$$

$$b = \frac{\Sigma xy \quad \dfrac{\Sigma x \; \Sigma y}{N}}{\Sigma x^2 - \dfrac{\Sigma (x)^2}{N}}$$

where x is the log of number of retail establishments, $\bar{x}$ is the mean of x, y is the log of retail functional index, $\bar{y}$ is the mean of y, N is the total number of retail cluster zones (22 in this case). Formula obtained from: Spiegel, M.R. (1961): "Theory and Problem of Statistics", Schaum's outline series, New York, p. 248.

11. Jones R. (1967): "Central Place Theory and the Hierarchy and Location of Shopping Centre in a city: Edinburg", cited by Carter, H. (1972): *The Study of Urban Geography*, Edward Arnold, London, p. 119.
12. King, A.M. and Cole, J.P. (1968): *Quantitative Geography*, London, p. 138.
13. Six socio-economic indicators are taken from Central Road Research Institute (1981): "Planning of Road System for Bombay Metropolitan Region", Vol. I Sec. II, pp. 9-16.
14. Census of India (1981): Primary Census Abstract, Part II B, Series 12, Maharashtra, pp. 250-252.
15. Sita, K. et al, (1985): "Declining City Care of an Indian Metropolis: A Case Study of Bombay, U.G.C. Project Report, Department of Geography, University of Bombay.
16. Clark C. (1951): "Urban Population Densities", *Journal of the Royal Statistical Society*, Sec. A, C XIV, Part IV, pp. 490-496.

7

CASE STUDIES OF RETAILING CLUSTERS IN GREATER BOMBAY

7.1 Selection of Zones for the Case Study

An attempt has been made in this chapter to make an in-depth case study of retailing characteristics of six retail cluster zones. The study was carried out with the help of a Questionnaire (Appendix III) related to various facets of retailing. For the purpose of the detailed case studies, six zones of Greater Bombay have been chosen which are representative of the different levels of retail hierarchy (barring the lowest order) and also fairly spread out and covering differing parts of the city and the suburbs. The six study areas are as follows.

TABLE 7.1: Case Study Areas

Area Chosen	Retail Hierarchy Level	Socio-Economic Level	Remarks
Fort	I	I	C.B.D.
Bhuleshwar-Kalbadevi	I	II	Commercial Core
Girgaum-Chowpatty	II	II	Old Residential Core
Dadar (West)	II	III	Early 20th Century Suburb, totally Residential
Bandra-Santacruz (W)	III	I	Inner-Western Suburb
Chembur-Kurla	III	IV	Inner-Eastern Suburb-mixed land-use with industry and residence

7.2 Sampling in the Study Areas

It has already been pointed out that the Bombay Municipal Corporation (B.M.C.) maintains a record of shops and establishments in a ward-wise (a ward is an administrative unit of the B.M.C.) manner, but there is no classification available from the B.M.C. records in terms of wholesale and retail establishments.

In these six zones selected for case studies, the number of registered shops in the B.M.C. record is 59,654, i.e., 51.4 per cent of the grand total of 1,35,545 registered shops in Greater Bombay. As against these 59,654 registered shops in the B.M.C. records, in the six selected zones, the number of shops enumerated by the researcher is 9,497, which is about 58 per cent of the total enumerated shops in all the 22 zones and 7 per cent of the total number of registered shops in Greater Bombay and 16 per cent of the registered shops in the six zones.

It may be noted that since the enumerated shops are only retail and in some cases semi-wholesale-cum-retail and since the B.M.C. record of registered shops does not separately classify retail and wholesale shops, the percentage share of the enumerated shops in these six zones (as well as in Greater Bombay, already explained in Chapter 6) with respect to the total registered shops will be much higher than 16 per cent and 7 per cent respectively, if retail and semi-wholesale shops only are considered for comparison. It will be reasonable to assume that these 9,497 enumerated retail and in some cases semi-wholesale shops will be roughly 20 per cent of the actual retail and semi-wholesale shops located in these six zones.

7.2.1 *The Sample Size*

A stratified, proportional and purposive sample of shops in each of the zones was undertaken as detailed below.

Of these 9,497 retail and semi-wholesale shops in the six case study areas, 526 shops under various commodity groups were surveyed with the help of a questionnaire, in depth. These 526 shops finally surveyed form nearly 5 per cent of the enumerated shops in the case of four zones located in the island city and about 9 per cent on an average of the enumerated shops in the case of the two remaining zones in the suburbs. This difference had to be effected in order that in smaller retail cluster zones of suburbs, most commodity groups could find representative members.

While conducting the survey with the questionnaire, it was experienced that the shopkeepers are extremely reluctant to respond, especially in regard to some sensitive aspects such as income or profit, because of fear of tax-raids. Hence, questions regarding financial status had to be kept out of the purview of the questionnaire. Thus, the use of retail sales as an economic indicator had to be ignored, except through indirect judgement of responses.

Associations of retailers, like the Grain-dealers Association, extended help to complete the survey. In spite of limitations and practical difficulties of completion, the survey produced worthwhile findings.

Table 7.2 shows the break-up of the sample in six zones and Table 7.3 shows the commodity-wise break-up.

TABLE 7.2: Share of Sample Shops in Relation to Registered Shops and Enumerated Shops in Each Zone

Sl. No.	Name of the Zone	Total No. of Registered Shops	Total No. of Enumerated Shops	Percentage Share of Enumerated Shops to Total Shops	Sample of Shops Surveyed	Percentage share of Sample Shops to Enumerated Shops
1.	Fort	6,929	2,673	38.6	132	5.0
2.	Bhuleshwar-Kalbadevi	14,120	2,748	19.5	137	5.0
3.	Girgaum-Chowpatty	12,209	1,306	10.7	65	5.0
4.	Dadar-(Plaza) Mahim	8,173	1,395	17.1	66	4.7
5.	Bandra-Santacruz (W)	5,419	568	10.5	50	8.8
6.	Chembur-Kurla	12,804	807	6.3	76	9.4
	Total :	59,654	9,497	15.9	526	5.5

Sources: 1. Bombay Municipal Corporatio;
2. Field Survey

Six maps, Maps 7.1 to 7.6 depicting the locations of the sample shops in the study areas were also plotted to facilitate spatial analysis.

The survey focuses on the following aspects:

(1) Age of shops
[shown in Table 7.4 (i-vii)]

(2) Religion-wise break-up of shopkeepers
[shown in Table 7.5 (i-vii)]

(3) Language group-wise break-up of shopkeepers
[shown in Table 7.6 (i-vii)],

TABLE 7.3: Distribution of Sample Shops in Each Zone by Types of Commodities

Name of the Zone	Total Sample	Electrical	Grocery	Textile	Sweetmeat-Milk	Footwear	Machinery	Utensiles Copper-brass	Tobacco Products	Variety Goods	Medicine	Paper	Jewellery	Sports goods toys	Crockery	Cold Storage Dry Fruits etc.	Vegetables	Sewing and Embroidary materials	Watch-Clock	Books & Periodicals	Photography-Zeroxing	Paints and Chemicals	Timber	Furniture	Bakery and Confectionary	Perfumes	Zari, glass miscellaneous
	1	2	3	4	5	6	7	8	9	10	11	12	13	14	15	16	17	18	19	20	21	22	23	24	25	26	27
1. Fort	132	10	9	40	4	—	14	—	—	2	—	16	—	—	6	6	—	—	4	—	—	—	—	—	—	—	21
2. Bhuleshwar Kalbadevi	137	10	—	42	—	11	8	8	—	—	7	11	10	—	—	—	—	—	—	—	—	—	—	—	—	—	30
3. Girgaum Chowpatty	65	5	-	19	-	-	13	7	-	-	-	5	5	—	—	—	—	—	—	—	—	—	—	—	—	—	10
4. Dadar (Plaza) Mahim	66	7	7	11	8	6	x	6	—	—	—	—	—	—	—	—	7	—	—	—	—	—	—	—	—	—	17
5. Bandra Santacuz (W)	50	–	6	10	—	6	—	—	—	8	—	—	—	—	—	—	—	—	—	—	—	—	—	—	—	—	20
6. Chembur Kurla	76	8	9	15	—	3	10	8	3	—	—	—	—	—	—	—	—	—	—	—	—	—	—	—	—	—	20

Source : Field Survey (1986).

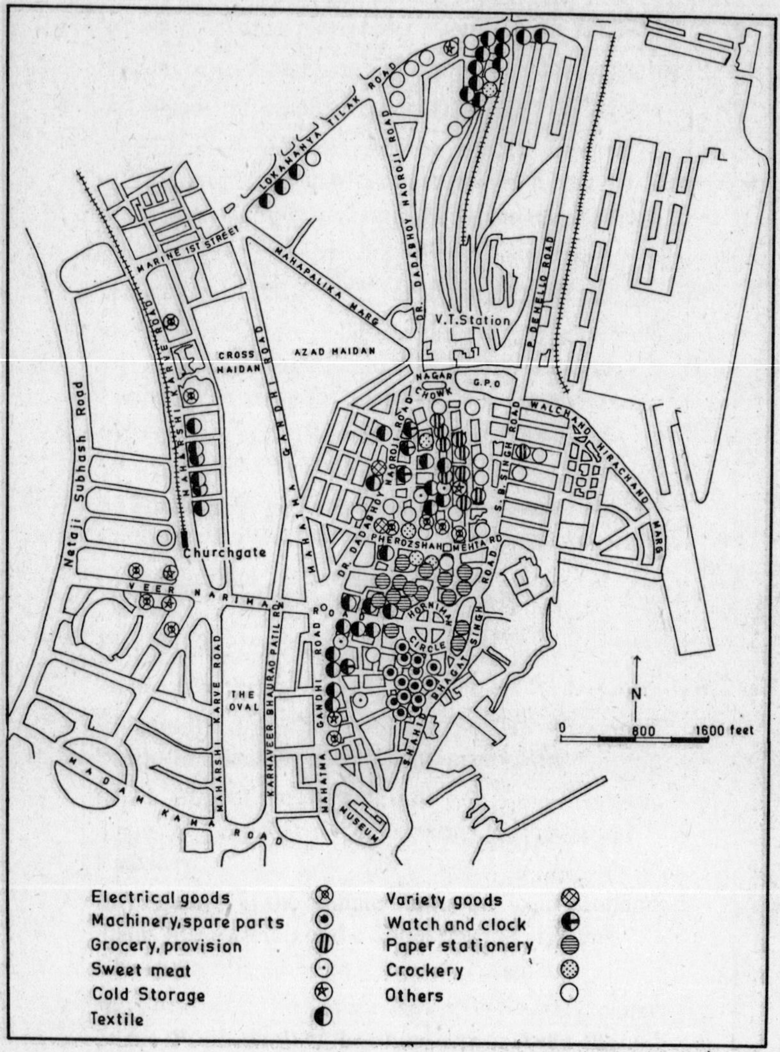

MAP 7.1: Fort Area (Shopping Sample)

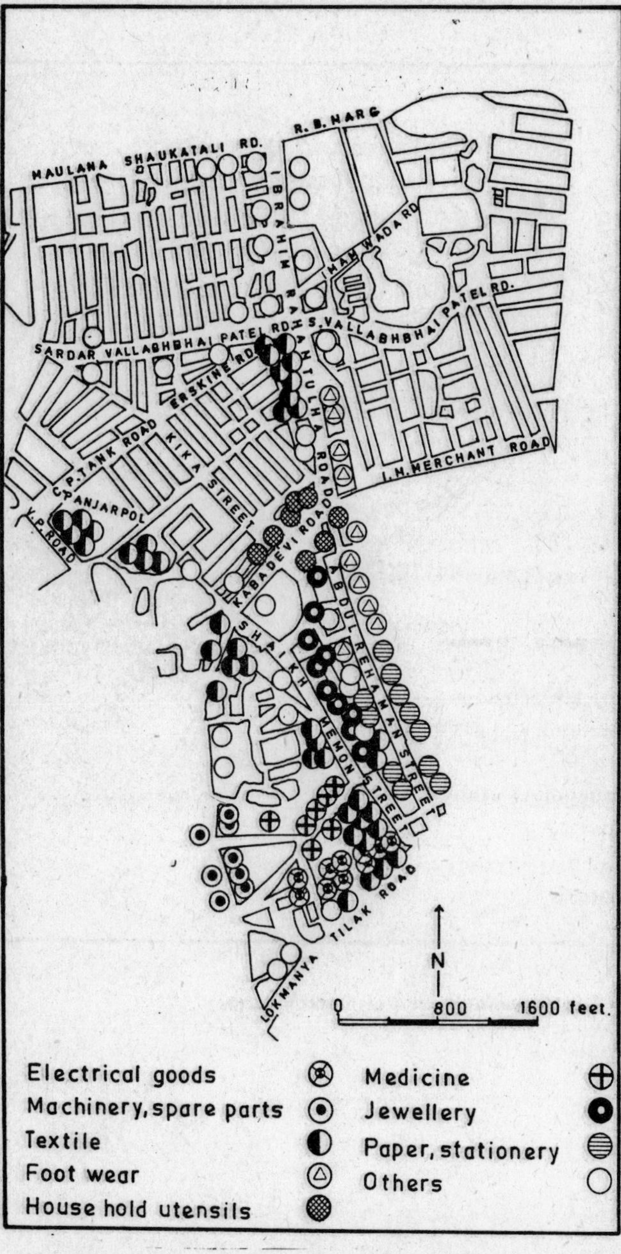

MAP 7.2: Bhuleswar-Market Area (Shopping Sample)

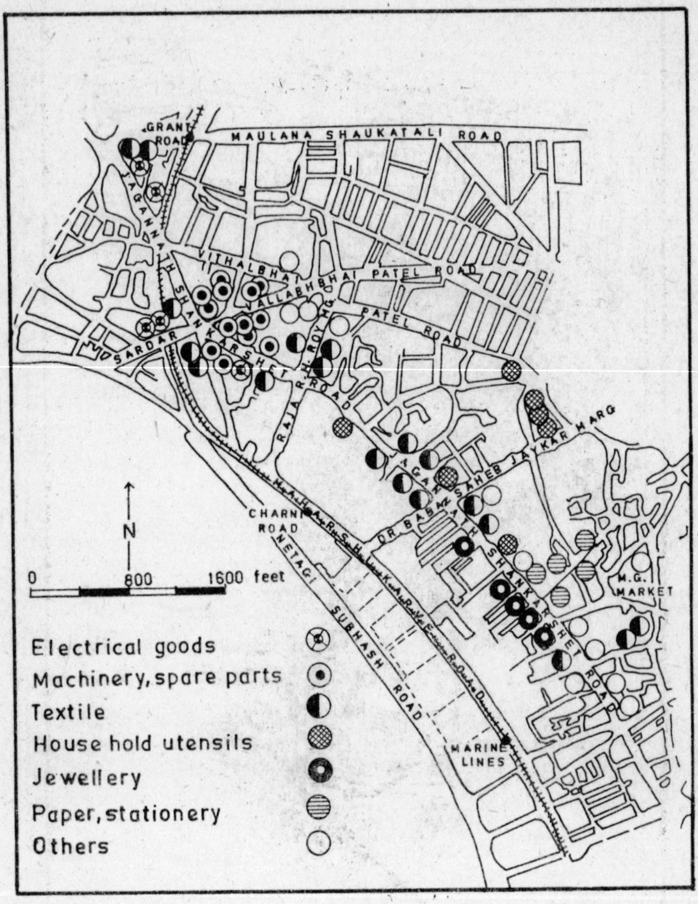

MAP 7.3: Girgaum Chaupati Area (Shopping Sample)

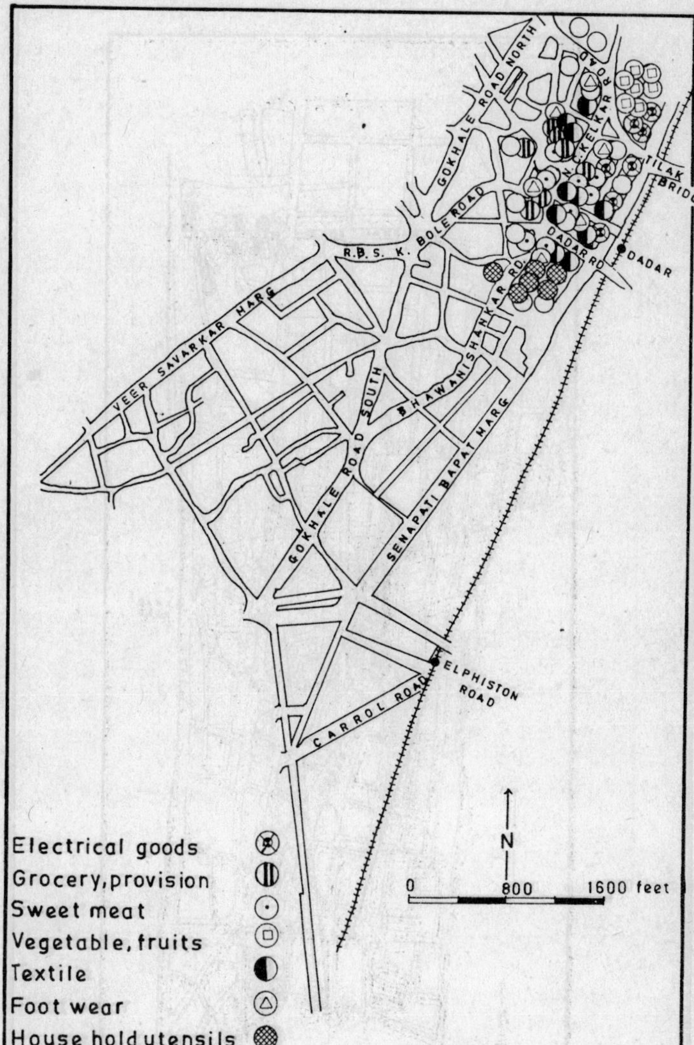

MAP 7.4: Dadar Area (Shopping Sample)

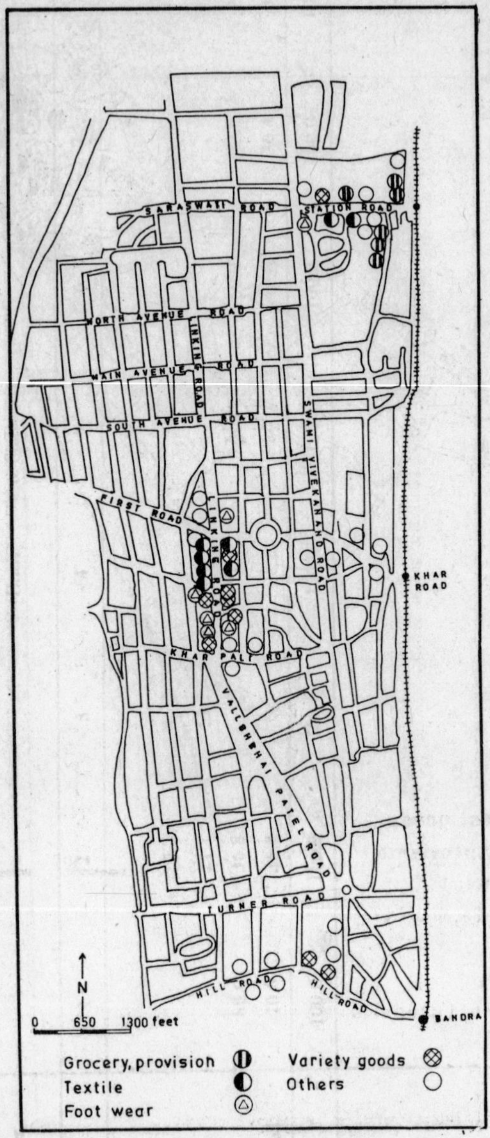

MAP 7.5: Bandra-Santacruz Area (Shopping Sample)

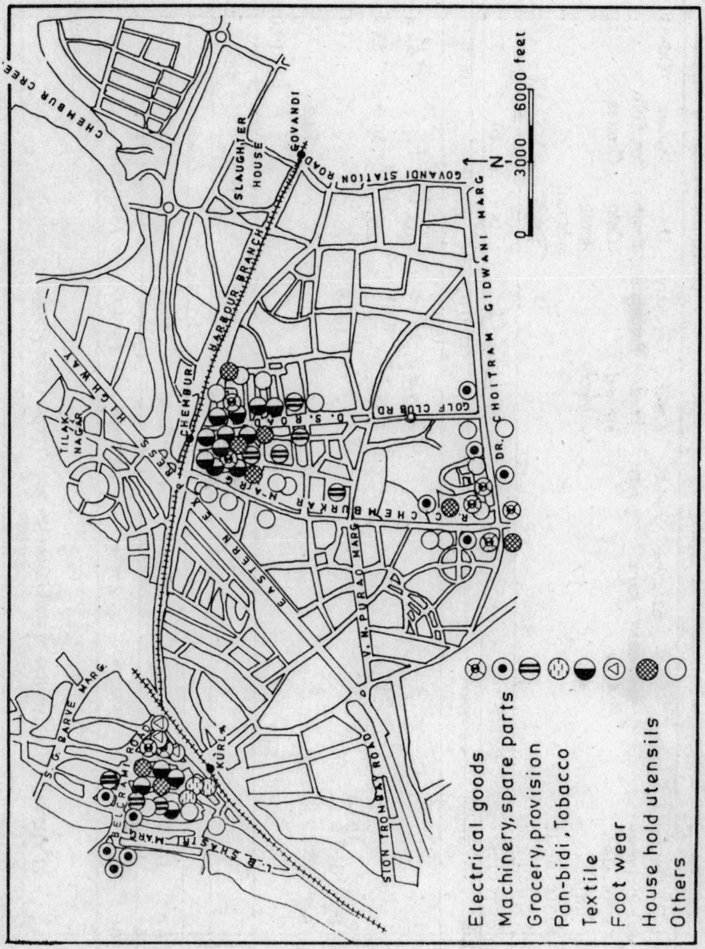

MAP 7.6: Kurla-Chembur Area (Shopping Sample)

TABLE 7.4: (i) Age of the Shops-Forts (Zone 2)

(Percentage Distribution)

Commodities Age of Shops (Years)	Electricals	Grocery Provisions	Textiles	Sweetmeat and Milk	Machine Parts	Variety goods	Paper, Printing and Stationery	Crockery Porcelain	Dry-Fruits Cold storage and Gen. Stores	Watch, Pen, Gift Centres	Others	Total
	1	2	3	4	5	6	7	8	9	10	11	12
Less than 5					7.1							0.75
5 to 10							12.5		33.3		23.8	6.8
11 to 25	40.0		25.0	100.0	78.6		43.8		66.7	50.0	23.8	35.62
26 to 40	30.0	100.0	50.0		7.1			100.0			33.4	34.8
41 to 55	30.0		20.0		7.1	50.00	6.3			50.0		12.1
56 to 70			5.0				18.8				19.0	6.8
71 to 90						50.00	6.3					1.5
Over 90							12.5					1.5
	100.0	100.0	100.0	100.0	100.0	100.0	100.0	100.0	100.0	100.0	100.0	100.0

TABLE 7.4: (ii) Age of Shops-Bhuleshwar-Kalbadevi (Zone 5).

(Percentage Distribution)

Commodities Age of Shops (Years)	Electricals	Textiles	Footwear Leather goods	Machine Parts	Utensils Copper & Brass	Medicine-Drugs	Paper Printing & Stationery	Jewellery	Others	Total
	1	2	3	4	5	6	7	8	9	10
Less than 5									6.7	1.5
5 to 10		4.8						10.0	6.7	2.9
11 to 25	10.0	23.8		12.5	12.5	28.6	27.3	10.0	16.7	17.5
26 to 40	70.0	52.4	27.3	50.0	50.0	71.4	45.5	50.0	50.0	48.2
41 to 55		14.3	72.7	37.5	37.5		27.3	30.0	20.0	24.8
56 to 70	20.0	4.8								5.1
71 to 90										
Over 90										
	100.0	100.0	100.0	100.0	100.0	100.0	100.0	100.0	100.0	100.0

TABLE 7.4: (iii) Age of the Shops, Girgaum-Chowpatty (Zone 8)

(Percentage Distribution)

Commodities Age of Shops Years	Electri-cals	Textiles	Machine Parts	Utensils Copper & Brass	Paper Printing & Stationery	Jewellery	Others	Total
	1	2	3	4	5	6	7	8
Less than 5			15.4		40.0	60.0		10.8
5 to 10		15.8	23.1		20.0			10.8
11 to 25	40.0	26.3	15.4		20.0		50.0	23.1
26 to 40	60.0	52.6	30.8	100.0	20.0	40.0	50.0	50.8
41 to 55		5.3	15.4					4.6
56 to 70								
71 to 90								
Over 90								
	100.0	100.0	100.0	100.0	100.0	100.0	100.0	100.0

TABLE 7.4: (iv) Age of Shops, Dadar (West) Plaza-Mahim (Zone 10)

(Percentage Distribution)

Commodities Age of Shops (Years)	Electri-cals	Grocery Provisions	Textiles	Sweet-meat & Milk	Footwear Leather Goods	Utensils Copper & Brass	Vegetables & Perishables	Others	Total
	1	2	3	4	5	6	7	8	9
Less than 5									
5 to 10				12.5		16.7		28.6	9.1
11 to 25	100.0	57.1	45.5	62.5	33.3	83.3		57.14	54.5
26 to 40		42.9	45.5	25.0	66.7		100.0	14.3	34.9
41 to 55			9.1						1.5
56 to 70									
71 to 90									
Over 90									
	100.0	100.0	100.0	100.0	100.0	100.0	100.0	100.0	100.0

Table 7.4: (v) Age of the Shops Bandra-Santacruz (West) (Zone 13)

(Percentage Distribution)

Commodities Age of Shops (Years)	Grocery- Provisions	Textiles	Footwear Leather Goods	Variety Goods	Others	Total
	1	2	3	4	5	6
Less than 5						
5 to 10	16.7	40.0	16.7	50.0	25.0	30.0
11 to 25	66.7	60.0	83.3	50.0	75.0	68.0
26 to 40	16.7					2.0
41 to 55						
56 to 70						
71 to 90						
Over 90						
	100.0	100.0	100.0	100.0	100.0	100.0

TABLE 7.4: (vi) Age of the Shops Kurla-Chembur (Zone 19)

(Percentage Distribution)

Commodities Age of Shops (Years)	Electri-cals	Grocery Provisions	Textiles	Footwear Leather Goods	Machine Parts	Utensils Copper & Brass	Tobacco & Products	Others	Total
	1	2	3	4	5	6	7	8	9
Less than 5								25.0	6.56
5 to 10	62.5	33.3	33.3	33.3	30.0	37.5	—	45.0	38.2
11 to 25	37.5	66.7	66.7	66.7	70.0	37.5	66.7	30.0	51.3
26 to 40						25.0	33.3		4.0
41 to 55									
56 to 70									
71 to 90									
More than 90									
	100.0	100.0	100.0	100.0	100.0	100.0	100.0	100.0	100.0

TABLE 7.4: (vii) Summary of Findings of the Case Studies in Six Zones of Bombay : Age of Shops

(Percentage Distribution)

Commodities Age of Shops (years)	Elec-tricals	Groc-ery Prov-isions	Text-iles	Sweet meat and Milk	Foot wear Leather Goods	Mach-ine Parts	Uten-sils, Copper & Brass	Toba-cco & Prod-ucts	Varie-ty Go-ods	Medi-cine Drugs	Paper, Printing & Stat-ionery	Jew-llery	Cro-ckery Porce-lain	Dry Fruits Cold Storage & Gen. Stores	Vege-tables Peri-shab-les	Others	Total
	1	2	3	4	5	6	7	8	9	10	11	12	13	14	15	16	17
Less than 5						6.7					6.3	20.0				6.0	2.9
5 to 10	12.5	12.9	10.2	8.3	7.7	13.3	13.8		40.0		9.4			33.3		18.1	13.3
11 to 25	42.5	45.2	33.6	75.0	34.6	46.7	31.0	66.7	40.0	28.6	34.4	6.7		66.7		34.5	37.1
26 to 40	32.5	41.9	41.6	16.7	26.9	20.0	44.8	33.3		71.4	18.8	20.0	100.0		100.0	31.0	32.7
41 to 55	7.5		11.7		30.8	13.3	10.3		10.0		12.5	33.3				6.9	10.3
56 to 70	5.5		2.9								9.4	20.0				3.5	3.0
71 to 90									10.0		3.1						0.4
Over 90											6.3						0.4
	100.0	100.0	100.0	100.0	100.0	100.0	100.0	100.0	100.0	100.0	100.0	100.0	100.0	100.0	100.0	100.0	100.0

(4) Size of retailing establishments measured in terms of number of employees
[shown in Table 7.7 (i-vii)]

(5) Average number of customers per day per shop under different categories of commodity groups
[shown in Table 7.8 (i-vii)], and finally

(6) Socio-economic (Residential) status of shopkeepers
[shown in Table 7.9 (i-vii)].

The information collected for the six select zones through the questionnaires was sorted and classified commodity-group-wise. The tabulated data were then analysed in terms of percentages, to get an insight of the socio-economic framework in which the retail economy functions in Greater Bombay.

7.3.0 *Main Findings of the Sample Study*

The results of this survey are now discussed in terms of the major retailing parameters and characteristics.

7.3.1 *The Age of the shop*

A systematic analysis of the age structure of the 526 sample retail shops in the six zones of Greater Bombay reveals the following striking features:

(i) Only four per cent of the shops are over 55 years old and less than 1 per cent over 70 years old. About 70 per cent shops are 11 to 40 years old. Barely 16 per cent are less than 10 years old; and

(ii) The oldest shops over 70 years are all in the Fort area, the original urban nucleus of the city, though old shops over 55 years are also found in Bhuleshwar and Fort. The two areas constitute together the original nucleus of wholesaling and the adjunct of retail cluster that developed with the first expansion of city beyond the Fort.

Girgaum-Chowpatty and Dadar areas have a predominance of medium-aged shops, while the suburban sample areas have a significantly larger proportion of younger shops.

Thus, on the whole there seems to exist an age gradient in retail shopping activity, from the city centre outwards towards the urban fringe:

the southern, older parts have a greater proportion of older shops, centre and north of the city are having middle-aged shops while the suburban areas are having younger shops predominating; possibly, the proportion of younger shops will be even higher in the outer suburbs. Thus, the proportion of age of shops shows a correspondence with the pattern of growth and sprawl of urban Bombay from the city core in the south outwards towards the north.

7.3.2 *Commodity-wise Distribution of Retail Shops*

Commodity-wise distribution of the age structure of the retail shops in the sample study shows that paper, printing and stationery shops are the oldest and they are exclusively concentrated in the Fort area. It is due to the fact that the entire publication industry is clustered in the Fort area even today. Till recently, all higher education facility was heavily clustered in that area. Roughly half the retail shops of paper and stationery are over 40 years old, with two main concentrations: the older shops in the city core and the relatively younger shops on its fringe in Girgaum-Chowpatty.

7.3.3 *Older Shops*

The textile retail market is moderately old. The older shops are located in South Bombay, while the younger ones are in North and Suburban Bombay. Almost all the other commodity retailing is young to moderately old. Almost all the crockery, porcelain shops, gift and watch-pen shops are 11 to 40 years old.

In case of most commodities, bulk of the retailing shops are 11 to 40 years old.

It is noted that in case of certain commodity groups there is very little later expansion and addition to the older shops and in some commodities there is a dynamism in the retail shopping characteristics, e.g. in the case of paper and stationery, textiles, machinery and electrical goods, temporal dynamism is visible, while in case of crockery, medicine and partly grocery in older areas, the age distribution is more or less stable.

7.4.1 *Affinity of Retail Shopkeepers with Religion*

Unlike the western context, in the Indian situation, retail trading is a family inheritance economic activity. Therefore, it is possible to identify strong social ties and affinities expressed through religions and languages in the Indian context. It is an empirical fact well-recognised not only in

Bombay but elsewhere in India that certain types of professions and trade activities go in strong association with specific communities. It is to examine the validity of this popular notion that this study was undertaken.

(i) Of the sample shops, approximately three-fourths are owned by Hindus and one-eighth by Muslims. Jains account for 7 per cent of ownership, Sikhs about 4.5 per cent and Parsees and Christians for less than 2 per cent each. It should be noted that the Parsee population is a mere 60,000 in the urban population of over 8 millions; hence, the share of Parsee retailing, though low, is quite significant for the size of the present population.

(ii) In all the sample zones, the predominance of ownership of Hindus is noted. However, it is strikingly high in Girgaum-Chowpatty, and Dadar-Mahim areas with almost 85 per cent under the ownership of Hindus. These are exclusively lower middle income residential areas of Hindu population, predominantly Maharashtrian. Chembur and Kurla areas also have got a higher percentage of Hindu shopkeepers than Bandra-Santacruz.

(iii) In contrast, Muslim ownership is restricted to specific zones. Bhuleshwar-Market has the highest concentration of Muslim shops, followed by Bandra-Santacruz and Kurla-Chembur. All these areas have sizeable pockets of Muslim resident population.

(iv) Jain ownership is predominant in Bhuleshwar, Girgaum Chowpatty and Dadar. They are found in all the sample survey zones, although their proportions vary significantly from area to area.

(v) Parsee ownership is almost exclusive to the Fort area, predominantly due to a long period of association almost from the beginning of the city. The Parsees are very old inhabitants of this area; moreover bulk of the Parsee resident population lies within a fringe of 2 to 4 kms from the Fort.

(vi) Christian ownership is primarily in Fort and Bandra-Santacruz. The latter area has a sizeable Christian resident population.

(vii) Sikh owership is found in all the areas except Dadar, with a heavier concentration in Kurla-Chembur and Girgaum-Chowpatty. The former area is close to the largest Sikh residential zone in the city, i.e., Koliwada-Sion. Thus, there appears a lingering association between the religious affinities of the retail consumers and shopkeepers, which is quite striking in Muslim, Parsee ownership.

7.4.2 *Commodity-wise Classification*

(i) The Hindu retail owners are spread over most of the commodities but some are almost exclusively in their hands; these are jewellery, medical shops, sweetmeats and utensils. They are the least in footwear and leather goods. They are relatively less active in machine parts, crockery and porcelain;
(ii) The Muslims have almost exclusive control over footwear and leather goods and account for a significant minority in textiles, machine parts, paper, electrical goods and variety goods.
(iii) Parsee ownership is primarily in crockery and secondarily in watch-pen and gift centre and to a small extent in paper.
(iv) Jain ownership dominates textiles, electrical goods, machine parts and grocery.
(v) The Sikhs predominate in machine parts, to a much lesser extent electrical goods and textile business and
(vi) Christian trading is mostly in minor commodities.

7.4.3 *Area-wise Classification*

Fort area has the most cosmopolitan shopkeeping, textiles and electrical goods being owned by Hindus and Muslims on a 2:1 proportion. Machine parts are shared between Hindus and Sikhs, crockery between Parsees, Hindus and Muslims. The rest of the commodity groups predominantly are in the hands of Hindus. In Bhuleshwar-Market area only Hindus, Jains and Muslims are the shopkeepers.

The Muslims are found in the eastern part adjoining the Mandvi-Chakala Wholesaling complex. Here the leather products are in the hands of Muslims, while jewellery and medicine are exclusively in the hands of Hindus; the rest are shared by them. But Muslims share a special association with some minor commodities like perfumes and bakery products.

Girgaum-Chowpatty is an exclusively Hindu shopping area. All commodities are handled by Hindus except for a share of the Sikhs in machine parts and the Jains in textiles and machine parts. The religious homogeneity of shopkeeping in this area is quite striking.

The Dadar shopping area is again homogeneous, 85 per cent of

TABLE 7.5(i) Religion of Shopkeepers Fort (Zone 2)

(Percentage Distribution)

Commodities Religion	Electricals	Grocery Provisions	Textiles	Sweetmeat & Milk	Machine Parts	Variety goods	Paper, Printing & Stationery	Crockery Porcelain	Dry-Fruits Cold storage & Gen. Stores	Watch, Pen, Gift Centres	Others	Total
	1	2	3	4	5	6	7	8	9	10	11	12
Hindu	70.0	100.0	67.5	100.0	42.9	100.0	68.8	33.3	83.3	75.0	47.6	65.2
Jain							12.5				19.0	4.5
Parsee							12.5	50.0		25.0	9.5	6.8
Sikh					57.1				16.7			6.1
Muslim	30.0		32.5				6.3	26.7				13.6
Christian											23.8	3.8
	100.0	100.0	100.0	100.0	100.0	100.0	100.0	100.0	100.0	100.0	100.0	100.0

TABLE 7.5: (ii) Religion of Shopkeepers, Bhuleshwar-Kalbadevi (Zone 5)

(Percentage Distribution)

Commodities Religion	Electricals	Textiles	Footwear Leather goods	Machine Parts	Utensils Copper & Brass	Medicine-Drugs	Paper Printing & Stationery	Jewellery	Others	Total
	1	2	3	4	5	6	7	8	9	10
Hindu	60.0	76.2		25.0	87.5	100.0	63.6	100.0	66.7	66.42
Muslim	30.0	16.7	100.0	37.5			18.2		16.7	22.6
Christian										
Parsee										
Jain	10.0	7.1		37.5	12.5		18.2		10.0	9.5
Sikh									6.7	1.5
	100.0	100.0	100.0	100.0	100.0	100.0	100.0	100.0	100.0	100.0

Table 7.5: (iii) Religion of Shopkeepers Girgaum-Chowpatty (Zone 8)

(Percentage Distribution)

Commodities Religion	Electri-cals	Textiles	Machine Parts	Utensils Copper & Brass	Paper Printing & Stationery	Jewellery	Others	Total
	1	2	3	4	5	6	7	8
Hindu	100.0	84.2	46.2	100.0	100.0	100.0	100.0	84.6
Muslim								
Christian								
Parsee								
Jain		15.8	15.4					7.7
Sikh			38.5					7.7
	100.0	100.0	100.0	100.0	100.0	100.0	100.0	100.0

TABLE 7.5: (iv) Religion of Shopkeepers—Dadar-Plaza-Mahim (Zone 10)

(Percentage Distribution)

Commodities Religion	Electri-cals	Grocery Provisions	Textiles	Sweet-meat & Milk	Footwear Leather Goods	Utensils Copper & Brass	Vegetables & Perishables	Others	Total
	1	2	3	4	5	6	7	8	9
Hindu	100.0	57.1	72.7	100.0	100.0	100.0	100.0	71.4	84.9
Muslim									
Christian									
Parsee									
Jain		42.9	27.3					28.6	15.2
Sikh									
	100.0	100.0	100.0	100.0	100.0	100.0	100.0	100.0	100.0

TABLE 7.5: (v) Religion of Shopkeepers, Bandra-Santacruz (West) (Zone 3)

(Percentage Distribution)

Commodities Religion	Grocery-Provisions	Textiles	Footwear Leather Goods	Variety Goods	Others	Total
	1	2	3	4	5	6
Hindu	100.0	80.0	33.3	75.0	65.0	70.0
Muslim			66.7	12.5	20.0	18.0
Christian					15.0	6.0
Parsee						
Jain		10.0				2.0
Sikh		10.0		12.5		4.0
	100.0	100.0	100.0	100.0	100.0	100.0

TABLE 7.5: (vi) Religion of Shopkeepers,Kurla-Chembur (Zone 19)

(Percentage Distribution)

Commodities / Religion	Electri-cals	Grocery Provisions	Textiles	Footwear Leather Goods	Machine Parts	Utensils Copper & Brass	Tobacco & Products	Others	Total
	1	2	3	4	5	6	7	8	9
Hindu	37.5	100.0	93.3	66.7		100.0	100.0	100.0	77.6
Muslim	25.0			33.3	60.0				11.8
Christian									
Parsee									
Jain	12.5								1.3
Sikh	25.0		6.7		40.0				9.2
	100.0	100.0	100.0	100.0	100.0	100.0	100.0	100.0	100.0

TABLE 7.5: (vii) Summary of Findings of the Case Studies in Six Zones of Bombay: Religion of Shopkeepers

(Percentage Distribution)

Commodities Religion	Electricals	Grocery Provisions	Textiles	Sweetmeat and Milk	Footwear Leather Goods	Machine Parts	Utensils, Copper & Brass	Tobacco & Products	Variety Goods	Medicine Drugs	Paper, Printing & Stationery	Jewellery	Crockery Porcelain	Dry Fruits Cold Storage & Gen. Stores	Vegetables Perishables	Watch Pen, Gift Centre	Others	Total
	1	2	3	4	5	6	7	8	9	10	11	12	13	14	15	16	17	18
Hindu	70.0	90.3	76.6	100.0	18.2	31.1	96.6	100.0	80.0	100.0	71.9	100.0	33.3	83.3	100.0	75.0	72.4	72.6
Muslim	20.0		14.6		84.6	20.0			10.0		9.4		16.6				7.8	12.7
Christian																	6.9	1.5
Parsee											6.3		50.0	16.7		25.0	1.7	1.7
Jain	5.0	9.7	7.3			11.1	3.5				12.5						9.5	6.8
Sikh	5.0		1.5			37.8			10.0								1.7	4.6
	100.0	100.0	100.0	100.0	100.0	100.0	100.0	100.0	100.0	100.0	100.0	100.0	100.0	100.0	100.0	100.0	100.0	100.0

shopping being in the hands of Hindus and the rest with Jains. Jains are predominant in case of grocery, textile and minor commodity groups.

Both in Bandra and Chembur-Kurla, retail shopkeepers are mixed, with a dominance of Hindus.

7.5.1 *Language Group-wise Break-up of Shopkeepers*

(i) Gujaratis constitute (36%) of retail ownership followed by the Kutchis (17%), Urdu-speaking Muslims (13%), Rajasthanis (9.5%), Maharashtrians (8.8%) and Sindhis (6.7%). Punjabis, South Indians and Hindi-speaking Uttar "Bharatiyas" account for abour 9 per cent. Thus, the retail shopkeepers of Bombay hail predominantly from Gujarat and Rajasthan. Local retailing talent is very limited.

(ii) Although the Fort region has a cosmopolitan shopkeeping character, Gujarati, Kutchi, Rajathani and Urdu-speaking shopkeepers among themselves account for three-fourths of ownership; Sindhis and Punjabis own one-eighth of the shops.

(iii) In Kalbadevi-Bhuleshwar, two-thirds of the shops are owned by Gujarati, Kutchi and Rajasthani people and almost a quarter by the Urdu-speaking Muslims. Marathi-speaking people have barely a 4 per cent ownership.

(iv) In Girgaum-Chowpatty, three-fourths of the shopkeepers originate from Gujarat and Rajasthan.

(v) Only in Dadar, Maharashtrians account for a third of the shopkeeping. Gujaratis and Kutchis even here own half of the shops.

(vi) In Kurla-Chembur, which is a low income area, Kutchis (28%) followed by Gujaratis (20%) and Rajasthanis (12%) are the main shopkeepers. Locally, here, the South Indians and Punjabis form significant minority owners each with 9 per cent. The Muslims have a 9 per cent ownership.

(vii) Bandra-Santacruz area has a mixed ownership pattern. Here, Sindhis constitute 20 per cent ownership, Kutchis and Rajasthanis together 38 per cent. Significantly, in this area, Gujarati ownership is very low, barely 6 per cent. Muslims account for 8 per cent.

7.5.2 *Commodity-wise break-up of Shopkeepers*

(i) Gujaratis almost exclusively control medical shops, tobacco shops, electricals, utensils, crockery, jewellery, watch and gift articles and dry fruits. They have significant ownership in machine parts, paper, textiles and variety goods.

(ii) Maharashtrians predominate in perishables, sweetmeat products, watch-pen and gift products, but have marked ownership in jewellery, footwear and sundry goods. Their ownership is relatively poor in textiles and electrical goods.

(iii) Sindhis predominate in variety goods and to a lesser extent in textiles (cutpieces) and footwear. Punjabis dominate in machine parts and to a lesser extent in variety goods. South Indians are seen only in variety goods. (Their presence in hotel-keeping is not covered in this study.) Hindi-speaking Uttar Bharatiyas dominate solely in sweetmeats. Urdu-speaking Muslims dominate in leather goods, machine parts and electrical goods.

(iv) Kutchis have an exclusive control over grocery and sizeable control over paper and dry fruits. They have sizeable ownership also in perishables and textiles. Rajasthanis have a share in textiles and jewellery.

7.5.3 *Areawise Break-up of Shopkeepers*

(i) In the Fort region, the Gujaratis are widespread in a number of commodities, but mainly in crockery, electrical goods, gift articles and dry fruits. Sindhis are prominent in watch-pen shops, Punjabis in machine parts, and Uttar Bharatiyas in sweetmeat; Kutchis are prominent in grocery, paper and dry fruits and Rajasthanis are prominent in textiles and electrical goods.

(ii) In Bhuleshwar-Kalbadevi, Gujaratis control a wide variety of commodities. They have got exclusive control over medicine and a sizeable control over utensils and electrical goods. Muslims have got exclusive control over footwear and leather goods and a sizeable control over electrical goods and machine parts. Sindhis have got prominent control over textiles (cutpieces), Rajathanis over textiles and paper, and Kutchis over paper.

(iii) In Girgaum-Chowpatty, Gujarati-speaking shopkeepers predominate in utensils, textiles and machine parts; Kutchis in paper, Rajasthanis in jewellery, Punjabis in machine parts, and Maharashtrians in utensils and textiles.

TABLE 7.6: (i) Mother Tongue of Shopkeepers Fort (Zone 2)

(Percentage Distribution)

Commodities Mother Tongue	Electricals	Grocery Provisions	Textiles	Sweetmeat and Milk	Machine Parts	Variety goods	Paper, Printing and Stationery	Crockery Porcelain	Dry-Fruits Cold storage and Gen. Stores	Watch, Pen, Gift Centres	Others	Total
	1	2	3	4	5	6	7	8	9	10	11	12
Gujarati	50.0		37.5		42.9	100.0	43.6	66.7	50.0	50.0	38.1	39.4
Marathi									16.7		9.5	2.3
Sindhi			12.5							50.0	9.5	6.8
Punjabi					57.1							6.1
South Indian Languages group											14.3	2.3
Hindi				100.0				16.7			9.5	5.3
Kutchi		100.0					56.3		33.3		19.1	18.2
Rajasthani	20.0		15.0									6.1
Urdu	30.0		35.0					16.7				13.6
	100.0	100.0	100.0	100.0	100.0	100.0	100.0	100.0	100.0	100.0	100.0	100.0

TABLE 7.6: (ii) Mother Tongue of Shopkeepers Bhuleshwar-Kalbadevi (Zone 5)

(Percentage Distribution)

Commodities Mother Tongue	Electricals	Textiles	Footwear Leather goods	Machine Parts	Utensils Copper & Brass	Medicine-Drugs	Paper Printing & Stationery	Jewellery	Others	Total
	1	2	3	4	5	6	7	8	9	10
Gujarathi	70.0	26.2		62.5	75.0	100.0	27.3	80.0	60.0	48.2
Marathi									16.7	3.7
Sindhi		9.5								2.9
Punjabi									6.7	1.5
Hindi		2.4								0.7
Urdu	30.0	14.3	100.0	37.5	25.0		18.2	10.00	16.7	23.4
Rajasthani		38.1					36.4	10.00		15.3
Kutchi							18.2			4.4
	100.0	100.0	100.0	100.0	100.0	100.0	100.0	100.0	100.0	100.0

TABLE 7.6: (iii) Mother Tongue of Shopkeepers, Girgaum-Chowpatty (Zone 8)

(Percentage Distribution)

Commodities Mother Tongue	Electri-cals	Textiles	Machine Parts	Utensils Copper & Brass	Paper Printing & Stationery	Jewellery	Others	Total
	1	2	3	4	5	6	7	8
Gujarati	80.0	63.2	61.5	85.7		60.0	36.4	52.3
Marathi		21.1		14.3			36.4	18.5
Sindhi	20.0							1.5
Punjabi			38.5					7.7
South Indian Group/								
Hindi/Urdu								
Kutchi		15.8			80.0		18.2	13.9
Rajasthani					20.0	40.0	9.1	6.2
	100.0	100.0	100.0	100.0	100.0	100.0	100.0	100.0

TABLE 7.6: (iv) Mother Tongue of Shopkeepers Dadar-Plaza-Mahim (Zone 10)

(Percentage Distribution)

Commodities Mother Tongue	Electricals	Grocery Provisions	Textiles	Sweetmeat & Milk	Footwear Leather Goods	Utensils Copper & Brass	Vegetables & Perishables	Others	Total
	1	2	3	4	5	6	7	8	9
Gujarati	71.4		45.5		16.7	50.0		21.4	25.8
Marathi	14.3		27.3	100.0	50.0	16.7	85.7	28.6	33.3
Sindhi	14.3				33.3				10.6
Parsee									
Punjabi									
South Indian Group/ Hindi/Urdu									
Kutchi		100.0	27.3			33.3	14.3	35.7	27.3
Rajasthani								14.3	3.0
	100.0	100.0	100.0	100.0	100.0	100.0	100.0	100.0	100.0

TABLE 7.6: (v) Mother Tongue of Shopkeepers, Bandra-Santacruz (West) (Zone 13)

(Percentage Distribution)

Commodities Mother Tongue	Grocery-Provisions	Textiles	Footwear Leather Goods	Variety Goods	Others	Total
	1	2	3	4	5	6
Gujarati		20.0		12.5		6.0
Marathi					15.0	6.0
Sindhi		30.0	33.3	37.5	10.0	20.0
Parsee						
Punjabi		10.0		12.5	10.0	8.0
South Indian Group				25.0		4.0
Urdu			66.7	12.5	20.0	18.0
Kutchi	100.0	20.0			25.0	26.0
Rajasthani		20.0			20.0	12.0
	100.0	100.0	100.0	100.0	100.0	100.0

TABLE 7.6: (vi) Mother Tongue of Shopkeepers Kurla-Chembur (Zone 19)

(Percentage Distribution)

Commodities Mother Tongue	Electri-cals	Grocery Provisions	Textiles	Footwear Leather Goods	Machine Parts	Utensils Copper & Brass	Tobacco & Products	Others	Total
	1	2	3	4	5	6	7	8	9
Gujarati	37.5		33.3			50.0	100.0	15.0	23.7
Marathi				33.3					1.3
Sindhi			6.7					15.0	5.3
Parsee									
Punjabi	25.0		6.7		40.0	37.5			9.2
South Indian Group	12.5							15.0	9.2
Hindi									
Urdu	25.0			33.3	60.0				11.8
Kutchi		100.0	20.0	33.3				40.0	27.6
Rajasthani			33.3			12.5		15.0	11.8
	100.0	100.0	100.0	100.0	100.0	100.0	100.0	100.0	100.0

TABLE 7.6: (vii) Summary of Findings of the Case Studies in Six Zones of Bombay: Mother Tongue of Shopkeepers

(Percentage Distribution)

Commodities Mother Tongue	Elec-tricals	Groc-ery Prov-isions	Text-iles	Sweet meat and Milk	Foot wear Leather Goods	Mach-ine Parts	Uten-sils, Copper & Brass	Toba-cco & Prod-ucts	Varie-ty Go-ods	Medi-cine Drugs	Paper, Printing & Stat-ionery	Jewe-llery	Cro-ckery Porce-lain	Dry Fruits Cold Storage & Gen. Stores	Vege-tables Peri-shab-les	Watch Pen, Gift Centre	Others	Total
	1	2	3	4	5	6	7	8	9	10	11	12	13	14	15	16	17	18
Gujarati	60.0		36.5		3.9	42.2	65.5	100.0	30.0	100.0	31.3	60.0	66.7	50.00		50.0	31.0	36.1
Marathi	5.0		5.1	66.7	15.4	6.9						20.0		16.7	85.7	50.0	12.1	8.8
Sindhi	2.5		9.5		15.4				30.0								9.5	6.7
Punjabi	5.0		1.5			37.8	10.3		10.0								3.4	4.9
South Indian Group	2.5								20.0								5.2	2.3
Hindi			0.7	33.3									16.7				1.7	1.5
Urdu	20.0		14.6		61.5	20.0	6.9		10.0		6.3		16.7				7.8	12.9
Kutchi		100.0	11.0		3.9		6.9				46.9			33.33	14.29		20.7	17.3
Rajasthani	5.0		21.2				3.5				15.6	20.0					8.6	9.5
	100.0	100.0	100.0	100.0	100.0	100.0	100.0	100.0	100.0	100.0	100.0	100.0	100.0	100.0	100.0	100.0	100.0	100.0

(iv) In Dadar-Mahim area, Gujaratis predominate in electrical goods, textiles and utensils. Maharashtrian control is prominent over sweetmeat, perishable goods, leather goods and partly on textiles and electrical goods. Sindhis have got predominant control over footwear and electrical goods. Kutchis have got exclusive control over grocery and sizeable control over utensils, textiles and perishable goods.

(v) In Kurla-Chembur, Gujaratis have got exclusive control over tobacco and a sizeable control over utensils and electrical goods. Maharashtrians are prominent in footwear and leather goods; Punjabis in machine parts and electricals; South Indians in electrical goods and Sindhis in readymade textile products; and Muslims mainly in machine parts and to a lesser extent in electrical and leather goods; Kutchis dominate grocery, footwear and textiles, and Rajasthanis textiles and utensils.

(vi) In Bandra-Santacruz, Kutchis have got exclusive control over grocery and sizeable control over textiles and some minor commodity groups. Sindhis mostly deal in textiles and footwear, and Muslims in leather goods.

(vii) On the whole, it may be noted that grocery seems to be exclusively in the hands of Kutchis, jewellery with the Gujaratis and Rajasthanis. Leather goods are with Muslims, machine parts with Punjabis and other products show mixed ownership.

7.6.0 *Economic Indicators*

While the first three indicators are social, the remaining indicators are economic. The number of employees per shop tells about the size. In the urban zone as a whole, 80 per cent of the retail shops are small with an employment of 6 or less and another 11 per cent is self/owner employment with no paid employee. Barely 9 per cent of the shops are large with a paid employee group of 7 and over.

7.6.1 *Number of Employees Per Shop*

Area-wise, Fort area has practically no self-employed shops. Small shops account for 86 per cent and the larger the rest.

In Bhuleshwar-Kalbadevi, 8 per cent of the shops are self-operated. 84 per cent belongs to the small category.

In contrast, in Girgaum-Chowpatty area as much as 31 per cent of the shops are operated by owners themselves, and 58 per cent are of small

TABLE 7.7: (i) Number of Employees Per Shop, Fort (Zone 2)

(Percentage Distribution)

Commodities No. of Persons Employed per Shop	Electricals	Grocery Provisions	Textiles	Sweetmeat & Milk	Machine Parts	Variety goods	Paper, Printing & Stationery	Crockery Porcelain	Dry-Fruits Cold storage & Gen. Stores	Watch, Pen, Gift Centres	Others	Total
	1	2	3	4	5	6	7	8	9	10	11	12
Over 10			2.5		14.3							2.3
9 to 7			10.0		35.7		6.3				23.8	11.4
6 to 4	30.0	66.7	32.5		35.7		31.3	33.3	85.0	75.0	47.6	39.4
3 to 1	70.0	33.3	52.5	100.0	14.3	100.0	62.5	66.7	15.0	25.0	28.6	46.2
Self-employed			2.5									0.8
	100.0	100.0	100.0	100.0	100.0	100.0	100.0	100.0	100.0	100.0	100.0	100.0

TABLE 7.7: (ii) Number of Employees Per Shop, Bhuleshwar-Kalbadevi (Zone 5)

(Percentage Distribution)

Commodities No. of Persons Employed per Shop	Electri-cals	Textiles	Footwear Leather goods	Machine Parts	Utensils Copper & Brass	Medicine-Drugs	Paper Printing & Station-ery	Jewellery	Others	Total
	1	2	3	4	5	6	7	8	9	10
Over 10		4.8						10.0		2.2
9 to 7	10.0	4.8		12.5	12.5			20.0		5.1
6 to 4	20.0	47.6	18.2	62.5	75.0	71.4	45.5	60.0	43.3	46.7
1 to 3	70.0	28.6	81.8	25.0	12.5	28.6	54.5	10.0	40.0	38.9
Self-employed		14.3							16.7	8.0
	100.0	100.0	100.0	100.0	100.0	100.0	100.0	100.0	100.0	100.0

TABLE 7.7: (iii) Number of Employees Per Shop Girgaum-Chowpatty (Zone 8)

(Percentage Distribution)

Commodities No. of persons Employed per Shop	Electri-cals	Textiles	Machine Parts	Utensils Copper & Brass	Paper Printing & Stationery	Jewellery	Others	Total
	1	2	3	4	5	6	7	8
Over 10								
9 to 7		5.3	46.2					10.8
6 to 4		21.1	53.9					16.9
1 to 3	80.0	36.8		71.4	20.0	60.0	63.6	41.5
Self-employed	20.0	36.8		28.6	80.0	40.0	36.4	30.8
	100.0	100.0	100.0	100.0	100.0	100.0	100.0	100.0

TABLE 7.7: (iv) Number of Employees Per Shop, Dadar-Plaza-Mahim (Zone 10)

(Percentage Distribution)

Commodities No. of persons Employed per Shop	Electri-cals	Grocery Provisions	Textiles	Sweet-meat & Milk	Footwear Leather Goods	Utensils Copper & Brass	Vegetables & Perishables	Others	Total
	1	2	3	4	5	6	7	8	9
More than 10									
9 to 7			9.1						1.51
6 to 4	28.6	100.0	18.2	37.5	16.7				22.73
3 to 1	57.1		63.6	62.5	33.3	83.3	71.4	71.4	57.6
Self-employed	14.3		9.1		50.0	16.7	28.6	28.6	18.2
	100.0	100.0	100.0	100.0	100.0	100.0	100.0	100.0	100.0

TABLE 7.7: (v) Number of Employees Per Shop, Bandra-Santacruz (West) (Zone 13)

(Percentage Distribution)

Commodities No. of persons Employed per Shop	Grocery-Provisions	Textiles	Footwear Leather Goods	Variety Goods	Others	Total
	1	2	3	4	5	6
More than 10		20.0				4.0
9 to 7		10.0			15.0	8.0
6 to 4	50.0	50.0	16.7	37.5	60.0	48.0
3 to 1	50.0	20.0	33.3	37.5	15.0	26.0
Self-employed			50.0	25.0	10.0	14.0
	100.0	100.0	100.0	100.0	100.0	100.0

TABLE 7.7: (vi) Number of Employees Per Shop, Kurla-Chembur (Zone 19)

(Percentage Distribution)

Commodities No. of persons Employed per Shop	Electri-cals	Grocery Provisions	Textiles	Footwear Leather Goods	Machine Parts	Utensils Copper & Brass	Tobacco & Products	Others	Total
	1	2	3	4	5	6	7	8	9
More than 10									
9 to 7			13.3	33.3	20.0				6.6
6 to 4	37.5	33.3	60.0		40.0	12.5		25.0	32.9
3 to 1	62.5	66.7	20.0	33.3	40.0	62.5	100.0	50.0	48.7
Self-employed			6.7	33.3		25.0		25.0	11.8
	100.0	100.0	100.0	100.0	100.0	100.0	100.0	100.0	100.0

TABLE 7.7: (vii) Summary of Findings of the Case Studies in Six Zones of Bombay : Number of Employees per Shop

(Percentage Distribution)

Commodities No. of Persons Employed per Shop	Electricals	Grocery Provisions	Textiles	Sweet meat and Milk	Foot wear Leather Goods	Machine Parts	Utensils, Copper & Brass	Tobacco & Products	Variety Goods	Medicine Drugs	Paper, Printing & Stationery	Jewellery	Crockery Porcelain	Dry Fruits Cold Storage & Gen. Stores	Vegetables Perishables	Watch Pen, Gift Centre	Others	Total
	1	2	3	4	5	6	7	8	9	10	11	12	13	14	15	16	17	18
More than 10			3.7		4.4								6.7					1.5
9 to 7	2.5		8.0		3.9	31.1	3.9					3.12	13.3				6.9	7.4
6 to 4	25.0	61.3	38.7	25.0	15.4	46.7	24.1		30.0		31.3	40.0	33.3	83.3		75.0	34.5	36.3
3 to 1	67.5	38.7	38.0	75.0	53.9	17.8	55.2	100.0	50.0	71.4	53.1	26.7	66.7	16.7	71.4	25.0	41.4	43.4
Self-employed	5.0		11.7		26.9		17.2		20.0	28.6	12.5	13.3			28.6		17.2	11.4
	100.0	100.0	100.0	100.0	100.0	100.0	100.0	100.0	100.0	100.0	100.0	100.0	100.0	100.0	100.0	100.0	100.0	100.0

size. In fact, there are not many shops with 10 employees or more. Girgaum stands out strikingly in the size composition.

In Dadar, 18 per cent of shops are operated by owners. Very large shops are not present in this area.

In the suburban zone, self-operated shops account for one eighth of the total shops and small shops constitute 65 to 80 per cent Bandra-Santacruz has a sizeable proportion (4%) of larger shops. It is pertinent to note the association in Girgaum and Dadar of the smaller self-operated shopkeeping with Maharashtrian Hindus and the larger shops with Gujaratis and Jains.

7.6.2 *Commodity-wise Larger Shops*

Commodity-wise, larger shops are mainly in jewellery, textiles and machine parts. Owner-operated shops are mainly in perishables, variety goods, footwear, utensils, paper and textiles.

It is to be noted that medium-size shops are mainly found in textiles, machine parts and jewellery. Almost all commodities have a predominant small-size establishments with 1 to 6 employees. The shop size in Fort area is small to medium because of high rental values and limited floor space available. Large to medium-size shops occur especially in jewellery, utensils, machine parts, textiles and electrical goods as found in Bhuleshwar-Kalbadevi, while small-size shopkeepers characterise the lower middle-income residential zone of Girgaum and Dadar, especially in utensils, textiles, electrical goods, jewellery, perishables and paper. The larger shops of Bandra and mainly textiles and of Chembur, textiles, machine parts and footwear. Gujarati Muslim ownership seems to have a greater degree of large size, while Maharashtrian and Uttar Bharatiya ownership are mainly self-operated and small-size establishments, that too in perishables, footwear and variety goods. Sindhi ownership also seems to be associated with self-operated establishments. Kutchi and Rajasthani owners in grocery and jewellery and textiles are mainly in small and medium-size shops.

7.7.1 *Number of Customer per day per Shop*

As per the number of customers per day, large-turnover shops, as measured by the size of customers catered to, account for about 11 per cent while 21 per cent have a small clientale (less than 25 customers per day). The rest fall in medium group where the number of customers ranges between 25 and 100.

TABLE 7.8: (i) Findings of the Sample Survey on Number of Customers Per Day Per Shop-Fort (Zone 2)

(Percentage Distribution)

Commodities No. of Customers per Day	Electri-cals	Grocery Provi-sions	Textiles	Sweet-meat & Milk	Machine Parts	Variety goods	Paper, Print-ing & Stationery	Crockery Porcelain	Dry-Fruits Cold storage & Gen. Stores	Watch, Pen, Gift Centres	Others	Total
	1	2	3	4	5	6	7	8	9	10	11	12
Less than 25	70.0		5.0		92.9		31.3			25.0		21.2
25 to 50	30.0	44.4	37.5		7.1	100.0	50.0	83.3	33.3		33.3	35.6
51 to 100		55.6	55.0	100.0			18.8	16.7	66.7	50.0	47.6	38.6
Over 100			2.5							25.0	19.1	4.5
	100.0	100.0	100.0	100.0	100.0	100.0	100.0	100.0	100.0	100.0	100.0	100.0

TABLE 7.8: (ii) Findings of the Sample Survey on Number of Customers Per Day Per Shop-Bhuleshwar-Kalbadevi (Zone 5)

(Percentage Distribution)

Commodities No. of Customers per Day	Electricals	Textiles	Footwear Leather goods	Machine Parts	Utensils Copper & Brass	Medicine-Drugs	Paper Printing & Stationery	Jewellery	Others	Total
	1	2	3	4	5	6	7	8	9	10
Less than 25		28.6	9.1		62.5			50.0		16.8
25 to 50	20.0	47.6	72.7	62.5	25.0	28.6	27.3	40.0	50.0	29.9
51 to 100	80.0	19.1	18.2	37.5	12.5	71.4	72.7	10.0	50.0	30.7
Over 100		4.8								22.6
	100.0	100.0	100.0	100.0	100.0	100.0	100.0	100.0	100.0	100.0

TABLE 7.8: (iii) Findings of the Sample Survey on Number of Customers Per Day Per Shop—Girgaum-Chowpatty (Zone 8)

(Percentage Distribution)

Commodities No. of Customers per Day	Electri-cals	Textiles	Machine Parts	Utensils Copper & Brass	Paper Printing & Stationery	Jewellery	Others	Total
	1	2	3	4	5	6	7	8
Less than 25	100.0	15.8	76.9	83.7	20.0	100.0	18.2	49.0
25 to 50		68.4	23.1		80.0		54.5	40.0
51 to 100		15.8		14.3			27.3	10.8
More than 100								
	100.0	100.0	100.0	100.0	100.0	100.0	100.0	100.0

TABLE 7.8: (iv) Findings of the Sample Survey on Number of Customers Per Day Per Shop Dadar-Plaza-Mahim (Zone 10)

(Percentage Distribution)

Commodities No. of Customers per Day	Electri-cals	Grocery Provisions	Textiles	Sweet-meat & Milk	Footwear Leather Goods	Utensils Copper & Brass	Vegetables & Perishables	Others	Total
	1	2	3	4	5	6	7	8	9
Less than 25	85.7							7.1	10.6
25 to 50	14.3		36.4		100.0	83.3		21.4	28.8
51 to 100		100.0	45.5	100.0		16.7		35.7	39.4
More than 100			18.2				100.0	35.7	21.2
	100.0	100.0	100.0	100.0	100.0	100.0	100.0	100.0	100.0

TABLE 7.8: (v) Findings of the Sample Survey on Number of Customers Per Day Per Shop, Bandra-Santacruz (West) (Zone 13)

(Percentage Distribution)

Commodities No. of Customers per Day	Grocery-Provisions	Textiles	Footwear Leather Goods	Variety Goods	Others	Total
	1	2	3	4	5	6
Less than 25		10.0			25.0	12.0
25 to 50		50.0	66.7	50.0	40.0	42.0
51 to 100	50.0	40.0	33.3	50.0	25.0	36.0
More than 100	50.0				10.0	10.0
	100.0	100.0	100.0	100.0	100.0	100.0

TABLE 7.8: (vi) Findings of the Sample Survey on Number of Customers Per Day Per Shop, Kurla-Chembur (Zone 19)

(Percentage Distribution)

Commodities No. of Customers per Day	Electricals	Grocery Provisions	Textiles	Footwear Leather Goods	Machine Parts	Utensils Copper & Brass	Tobacco & Products	Others	Total
	1	2	3	4	5	6	7	8	9
Less than 25	100.0			33.3		37.5		25.0	22.4
25 to 50			40.0		50.0	37.5		30.0	26.3
51 to 100		100.0	60.0	33.3	50.0	25.0		45.0	46.1
More than 100				33.3			100.0		5.3
	100.0	100.0	100.0	100.0	100.0	100.0	100.0	100.0	100.0

TABLE 7.8 (vii): Summary of Findings of the Case Studies in Six Zones of Bombay: Number of Customers Per Day Per Shop

(Percentage Distribution)

Commodities No. of Customers per Day	Electricals	Grocery Provisions	Textiles	Sweet meat and Milk	Foot wear Leather Goods	Machine Parts	Utensils, Copper & Brass	Tobacco & Products	Variety Goods	Medicine Drugs	Paper, Printing & Stationery	Jewellery	Crockery Porcelain	Dry Fruits Cold Storage & Gen. Stores	Vegetables Perishables	Watch Pen, Gift Centre	Others	Total
	1	2	3	4	5	6	7	8	9	10	11	12	13	14	15	16	17	18
Less than 25	65.0		13.1		7.7	51.1	48.3				18.8	66.7				25.0	11.2	21.5
25 to 50	15.0	12.9	46.0		69.2	31.1	34.5		60.0		37.5	26.7	83.3	33.3			25.9	33.1
51 to 100	20.0	77.4	37.2	100.0	19.2	17.8	13.8		40.0	28.6	18.8	6.7	16.7	66.7		50.0	32.8	34.0
Over 100		9.7	3.7		3.9		3.5	100.0		71.4					100.0	25.0	30.2	11.4
	100.0	100.0	100.0	100.0	100.0	100.0	100.0	100.0	100.0	100.0	100.0	100.0	100.0	100.0	100.0	100.0	100.0	100.0

Area-wise, larger clientale shops of over 100 per day are primarily in Bhuleshwar-Kalbadevi and Dadar-Mahim, while Girgaum-Chowpatty, Bandra-Santacruz also have about 10 per cent shops under this category.

A limited clinetele shop characterises the shopping complex of Girgaum-Chowpatty, where half of the shops belong to this class. This is true to a lesser extent of Kurla-Chembur.

7.7.2 Commodity-wise, the ubiquitous or near-ubiquitous products like perishables, tobacco, bakery products and medicine serve the maximum number of customers per day. The least clientele per day is seen in jewellery, electrical goods, machine parts and utensils that do not serve constant use of short-term replacement. Most commodities have moderate clientele. Grocery has a high clientele though not highest.

7.7.3 (i) Area-wise, in Fort the highest clientele is characteristically associated with watch, pen, gift shops mainly because it is the C.B.D., the administrative core with large service commuters, but partly because luxuries, commodities including smuggled goods are exclusively available in this area. Sweetmeat and milk products, textiles and dry fruits also have a moderately high clientele-all these being associated with the elite sort of commuting retailers who come to the area for specific purchases. Though the grocery clientale is sizeable, it is not as high as in other areas. The low clientele of less 25 per day in this area is mainly in machine parts, electrical goods and paper.

(ii) The Bhuleshwar-Kalbadevi area has a predominance of moderate to moderate high clientele in all the commodities, barring utensils which has low clientele. Here again, the largest clientele is in textiles.

(iii) In Girgaum area, most shops have small clientele. Fairly large clientele is in textiles and utensils and least in jewellery and electrical goods.

(iv) Dadar area being located in the intra-city transport focal point has got direct access to the northern as well as southern parts of the island. It has its largest clientele in perishables. Sweetmeats and grocery also have large clientele. Except electricals, all the shops have moderately heavy daily turnover.

(v) Bandra-Santacruz has its largest clientele in grocery, large in textiles, footwear and variety goods.

(vi) In Kurla-Chembur, the largest turnover is in tobacco and

TABLE 7.9 (i): Findings of the Sample Survey on Socio-Economic Status (Residential) of the Shopkeepers— Fort (Zone 2)

(Percentage Distribution)

Commodities Socio-economic Status Level	Electricals	Grocery Provisions	Textiles	Sweetmeat & Milk	Machine Parts	Variety goods	Paper, Printing & Stationery	Crockery Porcelain	Dry-Fruits Cold storage & Gen. Stores	Watch, Pen, Gift Centres	Others	Total
	1	2	3	4	5	6	7	8	9	10	11	12
I	30.0		50.0	75.0	100.0		25.0	33.3	83.3	75.0	23.8	42.4
II	30.0	100.0	10.0			100.0	56.3	33.3		25.0	4.8	24.2
III			5.0				6.3		16.7		47.6	12.1
IV	20.0											1.5
V	20.0		32.5	25.0			12.5	33.3			23.8	18.9
VI			2.5									0.8
	100.0	100.0	100.0	100.0	100.0	100.0	100.0	100.0	100.0	100.0	100.0	100.0

TABLE 7.9 (ii): Findings of the Sample Survey on Socio-Economic Status (Residential) of the Shopkeepers—Bhuleshwar-Kalbadevi (Zone 5)

(Percentage Distribution)

Commodities Socio-economic Status Level	Electri-cals	Textiles	Footwear Leather goods	Machine Parts	Utensils Copper & Brass	Medicine-Drugs	Paper Printing & Station-ery	Jewellery	Others	Total
	1	2	3	4	5	6	7	8	9	10
I	40.0	23.8	--	62.5		14.3	36.4	90.0		24.1
II		2.4								0.7
III		21.4			50.0					9.5
IV										
V	60.0	52.4	100.0	37.5	50.0	85.7	63.6	10.0	100.0	65.7
VI										
	100.0	100.0	100.0	100.0	100.0	100.0	100.0	100.0	100.0	100.0

TABLE 7.9 (iii): Findings of the Sample Survey on Socio-Economic Status (Residential) of the Shopkeepers—Girgaum-Chowpatty (Zone 8)

(Percentage Distribution)

Commodities Socio-economic Status Level	Electri-cals	Textiles	Machine Parts	Utensils Copper & Brass	Paper Printing & Stationery	Jewellery	Others	Total
	1	2	3	4	5	6	7	8
I		21.1	76.9	14.3				23.1
II								
III		5.3	23.1				27.3	10.8
IV								
V	100.0	73.7		85.7	100.0	100.0	72.7	66.2
	100.0	100.0	100.0	100.0	100.0	100.0	100.0	100.0

TABLE 7.9 (iv): Findings of the Sample Survey on Socio-Economic Status (Residential) of the Shopkeepers, Dadar-Plaza-Mahim (Zone 10)

(Percentage Distribution)

Commodities Socio-economic Status Level	Electri-cals	Grocery Provisions	Textiles	Sweet-meat & Milk	Footwear Leather Goods	Utensils Copper & Brass	Vegetables & Perishables	Others	Total
	1	2	3	4	5	6	7	8	9
I									
II									
III	14.3		18.2		16.7				6.1
IV	85.7	100.0	81.8	100.0	83.3	100.0	100.0	100.0	93.9
V									
VI									
	100.0	100.0	100.0	100.0	100.0	100.0	100.0	100.0	100.0

TABLE 7.9 (v) : Findings of the Sample Survey on Socio-Economic Status (Residential) of the Shopkeepers—Bandra-Santacruz (West) (Zone 13)

(Percentage Distribution)

Commodities Socio-economic Status Level	Grocery-Provisions	Textiles	Footwear Leather Goods	Variety Goods	Others	Total
	1	2	3	4	5	6
I	100.0	100.0		62.5	60.0	66.0
II						
III			100.0		40.0	28.0
IV				37.5		6.0
V						
VI						
	100.0	100.0	100.0	100.0	100.0	100.0

TABLE 7.9 (vi) : Findings of the Sample Survey on Socio-Economic Status (Residential) of the Shopkeepers Kurla-Chembur (Zone 19)

(Percentage Distribution)

Commodities Socio-economic Status-Level	Electri-cals	Grocery Provisions	Textiles	Footwear Leather Goods	Machine Parts	Utensils Copper & Brass	Tobacco & Products	Others	Total
	1	2	3	4	5	6	7	8	9
I									
II									
III	62.5							20.0	11.8
IV	37.5	100.0	100.0	100.0	100.0	100.0	100.0	80.0	88.2
V									
VI									
	100.0	100.0	100.0	100.0	100.0	100.0	100.0	100.0	100.0

TABLE 7.9 (vii) : Summary of Findings of the Case Studies in Six Zones of Bombay: Socio-Economic Status (Residential) of the Shopkeepers

(Percentage Distribution)

Commodities Socio-economic Status Level	Elec-tricals	Groc-ery Prov-isions	Text-iles	Sweet meat and Milk	Foot wear Leather Goods	Mach-ine Parts	Uten-sils, Copper & Brass	Toba-cco & Prod-ucts	Varie-ty Go-ods	Medi-cine Drugs	Paper, Printing & Stat-ionery	Jewe-llery	Cro-ckery Porce-lain	Dry Fruits Cold Storage & Gen. Stores	Vege-tables Peri-shab-les	Watch Pen, Gift Centre	Others	Total
	1	2	3	4	5	6	7	8	9	10	11	12	13	14	15	16	17	18
I	17.5	19.4	32.1			64.4	3.4		50.0	14.3	25.0	60.0	33.3	83.3		75.0	14.6	26.1
II	7.5	29.0	3.7	25.0							28.1		33.3			25.0	0.9	6.3
III	15.0		10.2		26.9	6.7	13.8		20.0		3.1						21.0	11.9
IV	27.5	51.6	17.5	66.7	30.8	22.2	48.2	100.0	30.0					16.7	100.0		25.9	25.5
V	32.5		35.8	8.3	42.3	6.7	34.5			85.7	43.8	40.0	33.3				37.0	30.0
VI			0.7															0.2
	100.0	100.0	100.0	100.0	100.0	100.0	100.0	100.0	100.0	100.0	100.0	100.0	100.0	100.0	100.0	100.0	100.0	100.0

bakery products, possibly associated with industrial working population. Footwear, leather goods, grocery, textiles and machine parts have a sizeably heavy clientele. Low clientele is found in electricals and to some extent utensils.

7.8.1 *The Residential Status of Shopkeepers*

The residential status of shopkeepers also reveals some interesting features.

About a third of the shopkeepers live in residential zones which can be ranked as upper and upper middle class residential areas. About 30 per cent live in low-income areas. The Fort retail shopkeepers seem to be associated with better class living; nearly two-thirds are living in upper class areas.

In sharp contrast, in Bhuleshwar and Girgaum areas, one-third live in upper class areas and two-thirds in middle income areas close to the shopping areas themselves.

On the other hand, the Dadar shopkeeper almost exclusively is local resident in lower middle-income residential areas.

Two-thirds of Bandra shopkeepers reside in upper class areas, while Kurla-Chembur has got mostly local residents in lower middle-income residential areas.

7.8.2 Commodity-wise the shopkeepers trading in jewellery, machine parts, dry fruits, cold storage, watch, pen and sizeable proportion of textiles, crockery and paper live in upper class areas. Shopkeepers dealing in tobacco, perishable products, bakery, sweetmeat, live in the lower income residential area. The rest of the commodities are widespread in middle-level income zones.

7.8.3 Area-wise Fort shopkeepers are resident in middle and upper class areas irrespective of the commodity group. Low-income residential living is mainly associated with local residents of Bhuleshwar and Girgaum areas, where the bulk of the shopkeepers are associated with shopping area itself, living in close proximity to the area itself in dilapidated housing. However, shopkeepers in jewellery, machine parts, paper and textiles live in upper class area. Dadar shopkeepers, irrespective of commodities, all belong to low-income resident group. In Bandra-Santacruz, shopkeeper of textiles, grocery and variety goods, live in upper brackets, Chembur-Kurla shopkeepers mainly belong to middle-income bracket.

7.9.1 *Procurement Source Survey*

An additional limited survey was also conducted on the nature of procurement of commodities. The survey was carried out in Bhuleshwar-Kalbadevi area, over a wide mix of commodities.

Though limited, this survey broadly indicates the nature of commodity flow pattern in the retail market. Procurement sources have been classified as wholesale market, factory, local dealer and agency, self-production and external sources.

It is noted that 43 per cent of the commodities of Bhuleshwar-Kalbadevi area are procured from wholesale market, 25 per cent from factory, only 15 per cent depend upon dealer and agency, 12 per cent are coming from outside sources and 4 per cent is self-produced (Table 7.10)

7.9.2 Commodity-wise, textile products are drawn directly from mill sources, which is close-by within the zone itself, namely, Mangaldas market and Mulji Jetha market.

Electrical goods are either from dealer or from factory, machine parts from factory or wholesale market, utensils solely from wholesale market which lies within the zone and jewellery entirely through the dealers. Medicinal products are primarily from dealers and to a lesser extent from wholesale market. Paper comes from wholesale market and partly from outside the city. Footwear exclusively depends on extra-city sources.

7.9.3 *In summary, it may now be stated that—*

(i) There exists an age-gradient in retail shopping activity, from the city centre outwards towards the urban fringe. The older clusters have a large concentration of older shops. Central and northern parts of the city have mostly middle-aged shops while suburban areas have relatively young shops.

(ii) Strong linguistic and religious affinities are seen in retail trading functions. The largest number of shops are owned by the Hindus. Muslims, Jains, Sikhs, Parsee and Christian owners have sizeable ownership of retail shops, quite out of relative proportion to their population in the metropolis.

(iii) There appears to exist a retailing differentiation according to religious and linguistic orders. Grocery is almost exclusively in the hands of Kutchis; jewellery in the hands of Rajasthani and Gujarati Hindus and Jains; leather goods with Urdu-

TABLE 7.10: Findings of the Sample Survey on Procurement of Commodities Bhuleshwar-Kalbadevi (Zone 5)

(Percentage Distribution)

Commodities Procurement from	Electricals	Textiles	Footwear Leather goods	Machine Parts	Utensils Copper & Brass	Medicine-Drugs	Paper Printing & Stationery	Jewellery	Others	Total
	1	2	3	4	5	6	7	8	9	10
Wholesale Market		52.4		37.5	100.0	28.6	63.6		62.5	43.5
Dealer	40.0	2.4				71.4		100.0		15.3
Factory	60.0	42.9		62.5					16.7	25.2
Outside City		2.4	100.0				36.4			12.2
Self-Manufacturer									20.8	3.8
	100.0	100.0	100.0	100.0	100.0	100.0	100.0	100.0	100.0	100.0

speaking Muslims; machine and auto-parts with Punjabis, and minor utility domestic products with Sindhis.

(iv) There also exists a shop size-commodity relation in urban retailing. Jewellery, textiles and machine parts are the larger ones. Owner-operated shops are mainly confined to perishables and variety goods.

(v) Ubiquitous or near-ubiquitous products like perishables, tobacco, bakery products and medicines serve the maximum number of customers per day. The least clientele shops are seen in electrical goods, machine parts and utensils.

(vi) Shopkeepers trading in jewellery, machine parts, dry fruits, cold storage, watch, pen and gift articles, crockery and a variety of textiles live in upper class areas. Shopkeepers dealing in tobacco, perishables, bakery and sweetmeat products live in lower income residential areas.

(vii) Fort shopkeepers are resident in middle and upper class areas irrespective of the commodity group. In Bhuleshwar and Girgaum areas, the bulk of shopkeepers are residentially associated with the shopping area itself. In case of Dadar, irrespective of commodities, shopkeepers belong to low-income residential group. In Bandra-Santacruz area, shopkeepers in textiles and variety goods live in upper class areas. Chemu-Kurla shopkeepers mainly belong to middle-income bracket.

8

THE INTERNAL STRUCTURE OF THE RETAIL CORE OF BOMBAY
A Case Study

8.1 Choice of Case Study Areas for Internal Morphology

Having examined the socio-economic structure of some selected retailing zones of Greater Bombay, it is now relevant to analyse the internal morphological structure of the retail core of the city in order to appreciate the physical landscape of retailing within the urban ecology of the city. This is done by analysing both the horizontal as well as vertical land-use patterns obtaining in two case study area, namely, the retail core of the Fort and Bhuleshwar-Kalbadevi which are commercially the most diversified and at the same time attracting even today a lion's share of retailing activity. Map 8.1 shows these two zones.

8.1.1 *The Fort Retail Core*

Zone one for the study is the Fort area that has the highest retailing intensity. Apart from public administration, it has a concentration of business houses, banks and financial institutions, entertainment places, and has the highest traffic interaction. The physical extent of the area is 92 hectares, stretching 800 metres north-south and 1,150 metres east-west. It corresponds to the B.M.C.'s administrative section, i.e., Fort North. It is a part of the retail cluster zone two. It is bounded by Esplanade Maidan in the west and north, Veer Nariman Road in the south and by the Docks in the east.

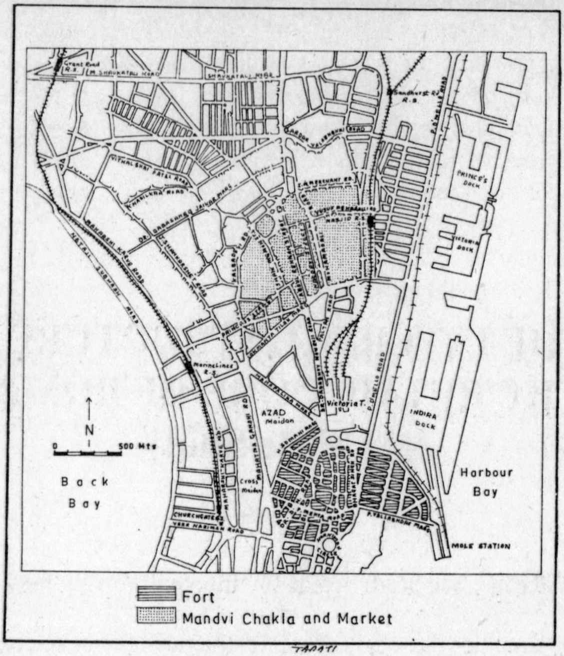

MAP 8.1: The Case Study Area

8.1.2 *Bhuleshwar-Kalbadevi-Mandvi Zone*

The second zone lies just north of the Fort comprising Bhuleshwar-Kalbadevi-Mandvi. It has a high wholesale intensity in its eastern parts, transforming into high retail intensity elsewhere in the zone.

It corresponds to parts of the B.M.C.'s administrative sections 6 and 7, Mandvi-Chakala, and section 13 of Bhuleshwar-Market. The area consists of parts of Zones 3 and 5 of retail cluster zones. It is bounded by Pydhonie Road in the north, Kalbadevi Road in the west, Lokmanya Tilak Road in the south and Harbour Railway Line in the east. It has an area of 75 hectares.

8.1.3 *Reasons for Choice of Areas*

The logic behind selecting this latter area combining parts of zones 3 and 5 of retail clusters for this detailed analysis is to show the spatial relations obtaining between wholesaling and retailing functions within the metropolitan core and to understand the nature of change-over of functions from wholesaling and semi-wholesaling to retailing across the area.

8.2 Methodology Adopted

Two plot-wise land use maps for each of these two zones, one for the ground floor and the other for the first floor, were made for the study of the spatial morphology and internal structure in these two selected zones. The study has been limited up to the first floor only because it was noted in the field that there is practically no retailing function in the upper storeys.

The map for each floor has been constructed based on a plot-to-plot and building-to-building field survey conducted during January to March, 1985. The working maps were plotted on a large scale of 1 cm=5 metres. For covering the Fort zone, 14 such maps had to be prepared. For Bhuleshwar-Kalbadevi 15 maps were made. The base maps were procured from the land records office of Government of Maharashtra in Bombay. The result of the survey was then transferred to two maps reduced on a scale of 1 cm = 48 m. (1" = 400¹), one for each zone for preparation of final maps.

For the purpose of mapping, these land-use functions have been grouped into (i) tertiary and (ii) non-tertiary functions and each sub-classified as shown in Table 8.1.

Based on the base maps, the total floor area of each zone on both ground and first floor exlcuding the roads and railway tracks, area under each function in each zone for both the floors and the areas under different functions alongside the main roads and lanes for both the zones for both the floors were determined by planimetric area calculations.

The reason for distinguishing functions alongside the main roads and inner roads and major lanes is to examine how micro-spatial variations in rentals and pedestrian as well as traffic flows influence the various functions.

8.2.1 *An American Parallel*

In case of a similar study in the U.S.A., Murphy (1954)[1] used built-up blocks for analysing the land-use functions. The way blocks have been planned in American cities to locate commercial function is not suitable for the present case study of Bombay. Main roads are the nerve-centres for Indian cities along which commercial function has grown, because they are the thoroughfares for pedestrian as well as vehicular traffic. Over time, these functions overflow into adjacent lanes and by-lanes as the main road fronts are fully occupied or because the rentals are too high for some retailing types to be afforded. Taking these aspects of oriental reality into

TABLE 8.1: The Broad Category of Land-Use Function Available in the Study Area

A. Tertiary		B. Non-Tertiary
1. Trading	2. Non-Trading	
(i) Retailing	(i) Banks	(i) Residence
(ii) Wholesaling	(ii) Commercial Establishments & Business Houses	(ii) Religious & Cultural Centres
(iii) Semi-Wholesaling	(iii) Govt. Services	(iii) Educational Institutions
(iv) Commission Agents		(iv) Recreational Centres
(v) Transport Services		(v) Light Industries
(vi) Godowns		(vi) Others

consideration, building plots are used as units in this study.

8.3 Internal Structure of Fort Retail Core

In Maps 8.2 and 8.3 various commercial as well as non-commercial functions have been plotted for the Fort area.

8.3.1 *Urban Land-Uses in Fort: A Gross View*

The proportionate percentage areas falling under each urban land-use in each floor in case of Fort North are shown in Table 8.2. The spatial distribution of the area under various tertiary and non-tertiary functions shown in Maps 8.2 and 8.3.

It is seen from Table 8.2 that retailing function in Fort is concentrated on the ground floor. Almost one-fourth (24.43%) of the total floor space on the ground floor is under retailing. On the other hand, only 1.96 per cent area is under retailing on the first floor. It is also noted that area-consuming trade functions such as wholesaling and godowns together form an insignificant proportion on both the floors. This is obviously related to the high rentals in the region and the keen rivalry among competing consumer and service-oriented land-uses for the prime land and street frontage on roads and corners on which pedestrian commuters' flow from all over the city for administrative and service employment converges and flows past. All available shopping fronts along the

TABLE 8.2: Internal Land Use Structure of the Fort Area

(All figures are in percentages of area under each function to the total area under all urban functions on each floor)

Functions/Services	Ground Floor %	First Floor %
1. Tertiary Services		
A. Trading		
(a) Retailing	24.43	1.96
(b) Non-Retailing		
(i) Wholesaling		
(ii) Semi-wholesaling		
(iii) Commission Agents		
(iv) Transport Services		
(v) Godowns	2.07	1.56
Sub-Total (A)	26.50	3.52
B. Non-Trading		
(i) Banks and Financial Institutions	9.62	10.08
(ii) Commercial Establishments and Business Houses	28.26	34.26
(iii) Govt. Offices	20.40	20.50
Sub-Total (B)	58.28	64.84
Total Tertiary Services (A+B)	84.78	68.36
2. Non-Tertiary (Other) Urban Functions:		
(i) Residence	1.76	18.24
(ii) Religious and Cultural Centre	2.37	1.83
(iii) Educational Institutions	4.28	3.98
(iv) Recreational Centre	3.93	2.92
(v) Light Industry	2.77	2.17
(vi) Others	0.10	2.50
Total Non-Tertiary Services	15.21	31.64
Grand Total (1 + 2)	100.00	100.00

Source : Data from Field Survey

commuting corridors are fully occupied by a wide variety of shopping and service functions. While wholesaling is practically absent, godowns are relegated to the interior niches, with lower rentals, that are not in the main commuter corridors.

Among non-trading functions, 9.62 per cent of the area is occupied by banks and financial institutions on the ground floor and slightly higher than this (10.08%) on the first floor. Commercial establishments and business houses occupy 28.26 per cent of the total area on the ground floor, the highest space occupied by any single function/service. More than one-third (34.26%) of the total area of the first floor again is occupied by business houses and commercial establishments. These establishments house the administrative functions of the commercial organisations and mail order function and do not require commodity handling. About a fifth of the total area on both the floors is occupied by government offices and this is related to the fact that Fort constitutes the old administrative core of the city, and inertia still retains a sizeable share of this function.

Tertiary services occupy 84.78 per cent of the space on the ground floor and 68.36 per cent on the first floor. As against this, non-tertiary functions occupy 15.21 per cent of total ground floor space and 31.64 per cent of first floor space. Residence take a large part of 18.24 per cent of area of the first floor. Residence, unable to compete with the high rent capacity of tertiary and production functions, occupy a very minor share (of 1.76 per cent of area) on the ground floor.

It is clear from this land-use study that retailing in Fort zone is predominantly on the ground floor, while residence and various other non-trading functions are located on the first floor and above. There is hardly any retail trading function on the first floor and where it exists, it is in commodities like cameras, radios and tape recorders that have a limited, elite clientele.

It is noted from field observations that in Fort area, the retailing core of the city, many retail shopkeepers carry on their commercial activities on the ground floor, while they reside with their families on the first floor, or on higher floors. This tradition is an old one that has been persisting since early days of the colonial city. However, as has already been mentioned, this tradition is steadily on the decline and the city core is declining in terms of residential population. A large number of traders who had been previously staying in the city core are now moving out to the suburbs or elsewhere in the city and are using or renting or selling the premises for trade storing, business office or other high rent activities because of high land price and or shortage of space in the city core.

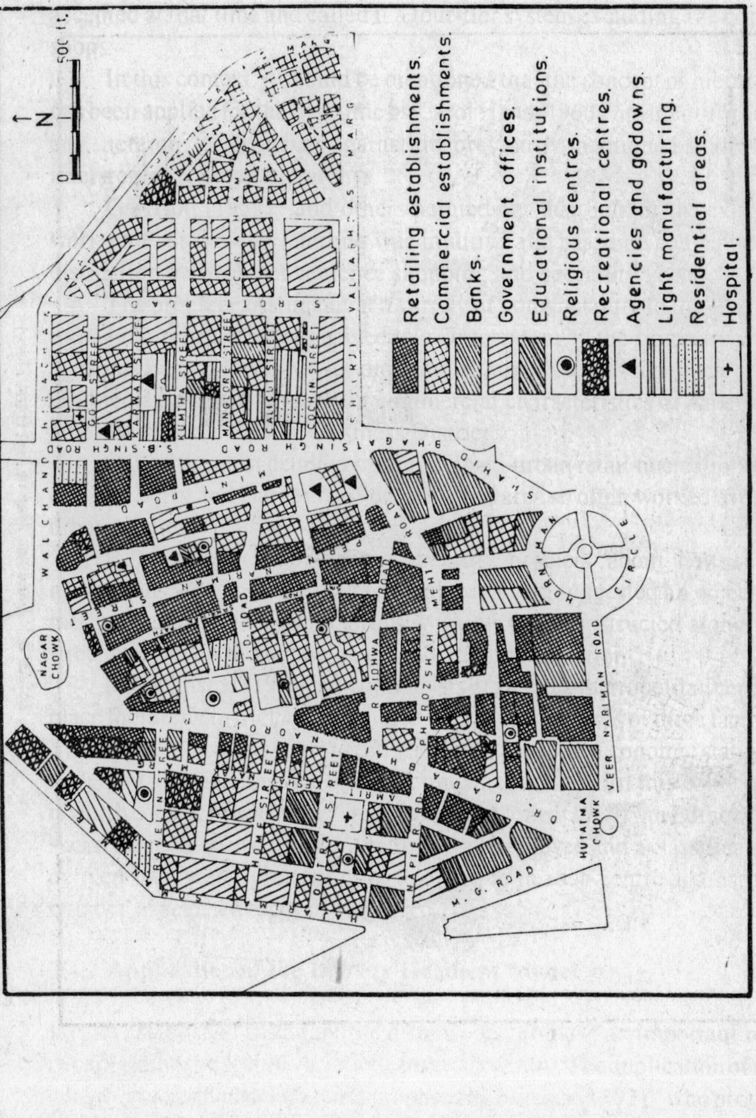

MAP 8.2: Land-use Map of Fort (Ground Floor)

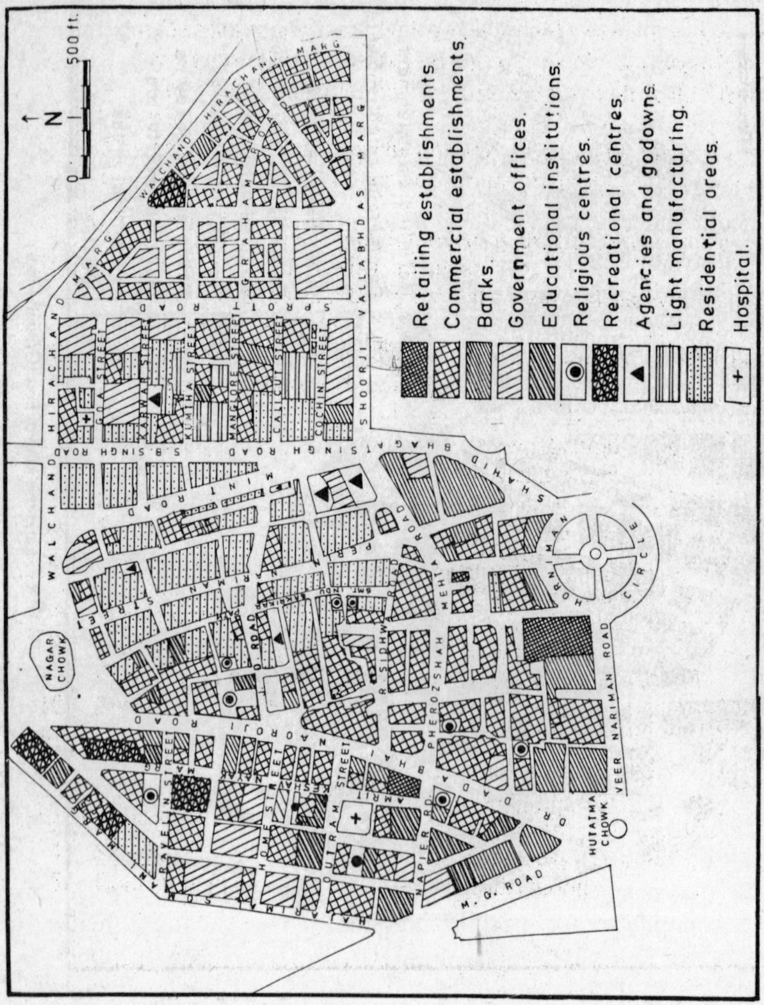

MAP 8.3: Land-use Map of Fort (First Floor)

8.3.2 *Road-wise Pattern of Retailing Distribution within the Fort Area*

A careful look at Maps 8.2 and 8.3 shows some salient features of the intra-zonal retail structure in the Fort area. Tables 8.3 and 8.4 show the road-wise pattern of land-uses among select roads in the Fort area, floor-wise.

Apart from the fact that the retailing function is almost exclusively on the ground floor, the internal areal differentiation is perceived in the quantum of distribution of retailing services, according as the road front is along the trunk roads, carrying the bulk of the commuter flow, or the side roads issuing into the main roads or by-lanes. This is seen not only in the total floor space occupied by retailing but also from the nature of commodity differentiation between different roads. This is illustrated through a study of 19 roads grouped into seven types. Of them, the Dadabhai Naoroji Road (D.N. Road), Bora Bazaar Street and Bazaar Gate Street (P. Nariman Road) are trunk roads in old Fort area carrying the main volume of pedestrian commuter flow, convergent on the rail-head terminal of Victoria Terminus, while the Murzban Road and Bastian Road are sub-parallel feeder roads to the main roads on areas developed after the demolition of the Fort wall with no vehicular/pedestrian flow.

It is also possible to distinguish between the retailing characteristics of the trunk roads like Dadabhai Naoroji Road (D.N. Road) and Shahid Bhagat Singh Marg and the inner roads like Bora Bazaar Street and Bazaar Gate Street which are located within the older dilapidated residential sections.

A minute look on Tables 8.3 and 8.4 shows that there is a high concentration of trading functions on the ground floor followed by non-trading functions like commercial establishments and business houses. The degree of concentration varies from one road to another. For example, in Bora Bazaar Street, 96 per cent of the total area is under retailing, while in D.N. Road, it is 63 per cent though D.N. Road is the main business thoroughfare. The lesser percentage in D.N. Road is not because of lesser retailing importance but because of a stiffer rivalry from other commercial and service functions on this trunk artery for street frontage.

It is apparent from Map 8.1 that D.N. Road and Shahid Bhagat Singh Marg (former D'Mello Road) are two trunk roads with vehicular traffic and pedestrian commuter flows; of them, D.N. Road is the main business thoroughfare, while the latter is peripheral in location along the eastern edge of the original fort, adjoining Mody Back Reclamations, where large business houses and some port functions are located.

To the west of D.N. Road secondary feeder roads like Napier Road, Purshottamdas Thakurdas Road, Outram Road, and Ravelin Street house

TABLE 8.3: Road-wise Distribution of Land-use Functions within Fort Area Zone-I - Ground Floor

(in percentage)

	Retailing	Wholesaling	Semi-Wholesaling	Commission Agent	Transport Agent	Godowns	Bank & Financial Institutions	Commercial Establishment & Business House	Govt. Office	Religious & Cultural Centre	Educational Institute	Recreation Centre	Light Industry	Residence	Others
	(1)	(2)	(3)	(4)	(5)	(6)	(7)	(8)	(9)	(10)	(11)	(12)	(13)	(14)	(15)
I. Main Road:															
(a) D.N. Road	62.98	—	—	—	—	—	11.08	6.62	—	11.04	—	8.28	—	—	—
(b) Shahid Bhagat Singh Rd.	26.85	—	—	—	—	11.11	23.14	4.20	20.37	—	—	—	13.88	—	—
II. Secondary Road West of D.N. Rd.															
(a) Bastian Rd.	—	—	—	—	—	—	—	25.55	4.10	2.70	36.66	26.66	—	—	5.2
(b) Murzban St.	—	—	—	—	—	—	5.20	27.40	25.30	2.70	23.10	13.20	—	3.90	—
III. Secondary Road East of D.N.Rd.:															
(a) Bora Bazaar Street	96.44	—	—	—	—	—	—	—	—	—	3.10	8.45	—	—	—
(b) Bazaar Gate Street	70.83	—	—	—	—	—	27.16	—	2.10	—	—	—	—	—	—

(Contd.)

Table 8.3: (Contd.)

	(1)	(2)	(3)	(4)	(5)	(6)	(7)	(8)	(9)	(10)	(11)	(12)	(13)	(14)	(15)
IV. Secondary Road on Both Sides of Bhagat Singh Rd.:															
(a) Mody Street	59.50	—	—	—	—	21.60	13.50	3.40	—	—	—	2.00	—	—	—
(b) Nicol Road	—	—	—	—	—	—	—	36.63	63.36	—	—	—	—	—	—
(c) Sports Road	—	—	—	—	—	—	—	46.45	38.46	—	—	15.08	—	—	—
V. Lanes on West of D.N. Road:															
(a) Napier Road	—	—	—	—	—	—	—	—	40.00	—	45.00	—	—	—	15.00
(b) Outram Road	—	—	—	—	—	—	—	42.00	—	—	47.36	—	—	—	10.52 (Hospital)
(c) Ravelin Street	—	—	—	—	—	—	9.09	22.72	—	22.70	9.09	22.72	—	13.63	
VI. Lanes on East of D.N. Road:															
(a) Ghoga Street	79.54	—	—	—	—	—	—	—	9.09	9.09	—	11.36	—	—	—
(b) Gunbow St.	75.00	—	—	—	—	—	—	25.00	—	—	—	—	—	—	—
(c) Phirozeshah Mehta Road	56.81	—	—	—	—	—	29.54	13.63	—	—	—	—	—	—	—
VII. Lanes on Both Sides of Bhagat Singh Marg:															
(a) Mint Road	50.00	—	—	—	—	27.77	—	12.96	9.25	—	—	—	—	—	—
(b) Cochin St.	7.69	—	—	—	—	—	7.69	15.38	37.61	—	—	—	23.07	7.69	—
(c) Calicut St.	15.38	—	—	—	—	—	7.69	38.46	—	—	—	—	30.76	7.69	—
(d) Mangalore St.	—	—	—	—	—	—	15.38	50.00	—	—	—	7.69	26.92	—	—

TABLE 8.4: Road-Wise Distribution of Land-use Functions within Fort Area Zone-I - First Floor

(in percentage)

	Retailing	Wholesaling	Semi-Wholesaling	Commission Agent	Transport Agent	Godowns	Bank & Financial Institutions	Commercial Establishment & Business House	Govt. Office	Religious & Cultural Centre	Educational Institute	Recreation Centre	Light Industry	Residence	Others
	(1)	(2)	(3)	(4)	(5)	(6)	(7)	(8)	(9)	(10)	(11)	(12)	(13)	(14)	(15)
I. Main Road:															
(a) D.N. Road	—	—	—	—	—	—	22.50	58.33	4.16	8.33	—	6.66	—	—	—
(b) Shahid Bhagat Singh Rd.	—	—	—	—	—	—	22.93	12.84	20.18	—	—	—	7.33	36.69	—
II. Secondary Road West of D.N. Rd.															
(a) Bastian Rd.	—	—	—	—	—	—	—	25.55	4.10	2.70	36.66	26.66	—	—	5.2 (Hospital)
(b) Murzban St.	—	—	—	—	—	—	—	27.40	25.30	2.70	23.10	13.20	—	3.90	—
III. Secondary Road East of D.N.Rd.:															
(a) Bora Bazaar Street	—	—	—	—	—	—	—	11.22	—	2.22	2.00	—	—	85.70	—
(b) Bazaar Gate Street	—	—	—	—	—	—	25.30	19.20	—	—	—	—	—	55.50	—

(Contd.)

Table 8.4: (Contd.)

	(1)	(2)	(3)	(4)	(5)	(6)	(7)	(8)	(9)	(10)	(11)	(12)	(13)	(14)	(15)
IV. Secondary Road on Both Sides of Bhagat Singh Rd.:															
(a) Mody Street	—	—	11.60	—	—	15.90	14.50	—	—	—	—	—	—	58.00	—
(b) Nicol Road	—	—	—	—	—	—	—	36.69	63.36	—	—	—	—	—	—
(c) Sports Road	—	—	—	—	—	—	—	46.45	38.46	—	—	15.08	—	—	—
V. Lanes on West of D.N. Road:															
(a) Napier Road	—	—	—	—	—	—	—	—	40.00	—	45.00	—	—	—	15.00
(b) Outram Road	—	—	—	—	—	—	—	42.00	—	—	47.36	—	—	—	10.52
(c) Ravelin Street	—	—	—	—	—	—	9.09	22.72	—	22.70	9.09	22.72	—	13.63	
VI. Lanes on East of D.N. Road:															
(a) Ghoga Street	—	—	—	—	—	—	—	64.00	16.00	—	—	—	—	20.00	—
(b) Gunbow St.	—	—	—	—	—	—	—	—	—	—	—	—	—	—	—
(c) Phirozeshah Mehta Road	6.15	—	—	—	—	—	30.76	58.46	—	4.61	—	—	—	—	—
VII. Lanes on Both Sides of Bhagat Singh Marg:															
(a) Mint Road	—	—	—	—	—	27.77	—	10.76	—	—	—	—	—	61.53	—
(b) Cochin St.	—	—	—	—	—	—	7.69	15.38	37.61	—	—	—	20.07	19.20	—
(c) Calicut St.	—	—	—	—	—	—	7.69	38.46	—	—	—	—	24.76	29.07	—
(d) Mangalore St.	—	—	—	—	—	—	15.38	50.00	—	—	—	7.69	26.92	—	—

educational institutions and business houses and recreational centres. 45 per cent of the total space of Napier Road is occupied by educational institutions and 40 per cent of the space of the same road is occupied by administrative offices. So also in Purshottamdas Thakurdas Road, 42 per cent of the space is occupied by business houses. This shows that these feeder roads contrast in their character with those on the other side of the D.N. Road (between D.N. Road and Shahid Bhagat Singh Marg) in the old Fort area, namely, Bora Bazaar Street and P. Nariman Road (Bazaar Gate Street).

In Bora Bazaar Street, the predominating function on ground floor is retailing, occupying 96.44 per cent of the total area. In P. Nariman Road (Bazaar Gate Street), 70.83 per cent of the total area is occupied by retailing and 27 per cent of the total area is occupied by banks and financial institutions. In the feeder roads like Ghoga Street, Gunbow Street and Phirozshah Mehta Road, that lead to the secondary service roads of Bazaar Gate and Bora Bazaar Streets, the predominating function is retailing. In the three roads, Ghoga Street, Gunbow Street and Phirozshah Mehta Road, the total area occupied by retailing is 79, 75 per cent and 56 per cent respectively.

In Phirozshah Mehta Road, the next important function is banking, in fact, it is part of the banking district of the city. This functional arrangement and its association on both sides of the D.N. Road, and the Shahid Bhagat Singh Marg can be explained by the fact that the road cuts through the old core of the Fort city, west to east, and is centrally poised for servicing the retailing function, commercial establishments, business houses, administrative offices and port functions.

In Bhagat Singh Marg, ground floor land-use is mixed, no function being excessively dominant. Retailing has the largest area occupation (27%), that is marginally higher than area occupied by banks and financial institutions followed by Government offices (20%), light industry (14%) and godowns (11%).

The secondary roads on both sides of Bhagat Singh Road, namely, Mody Street, Nicol Road and Sprott Road, also have a mixed functional land-use, but Mody Street has 60 per cent of the area occupied by retailing, 22 per cent by godowns and 14 per cent by the banks. In Nicol Road and Sprott Road, retailing is conspicous by its absence, the predominant function being business houses and Government offices. Of the lesser important roads around, the Mint Road has 50 per cent of the total area occupied by retailing, 28 per cent by godowns, 13 per cent by commercial establishments, while Cochin Street and Calicut Street have barely 8 per cent and 15 per cent under retailing, the rest being under administration and light industry.

This contrast between Mody Street and others is obviously due to the fact that Mody Street is closer in location to the retail core of Bazaar Gate Street (P. Nariman Road) and Bora Bazaar Street, while the others are totally peripheral in their location to commuter flows.

In contrast to the ground floor's variety of urban land-use functions, the first floor in Fort has practically a total absence of trading function, especially retailing. In case of D.N. Road, the prime space on first floor is occupied by business houses followed by banks and financial institutions. In case of Shahid Bhagat Singh Marg, the dominant function on the first floor is residential accommodation (36.69%) followed by banks and financial institutions.

On the roads to the west of D.N. Road, namely, Murzban Street and Bastian Road, two-thirds of the space on the first floor is occupied by business houses and educational institutions. Both trading and residential functions are totally absent. This can be again explained by the fact that in these later occupied areas, because of higher land values and planned European architectural styles, the space allocation shows the rent paying capability of the users. Development of this land-use zone was also pampered by the early European and Parsee residential quarters located adjacent to it. In contrast to the eastern side of D.N. Road, the Bora Bazaar Street and P. Nariman Street (Bazaar Gate Street) have residence as the predominant function, 86 per cent and 56 per cent respectively. Most of the dwellers here are retail traders of the zone who have been settled here since the inception of the European fortified township. In fact, this residential core of old Fort area is a fossilised function that has withstood all onslaughts of the subsequent decades by varied urban functions, and this is rooted in the social, community affinities to specific retail trade functions of the area, discussed in earlier chapters.

Eastward beyond Mody Street, the pattern gradually changes. Mody Street and Mint Road are part of the old core where 58 per cent of the total area is occupied by residence. But further east, on Nicol Road and Sprott Road and the by-lanes, prime space is occupied by business houses, government offices and light industry.

On the whole, in the section between D.N. Road and Mody Street lies the main retailing core of the Fort, where the retailing function is the dominant land-use function on the ground floor. This is also the residential core of the native population within the Fort, the settlement of the trading community. It slightly declines in importance from the main roads to the secondary roads and feeder roads. The three roads on which commuter flow to and from the Victoria Terminus (V.T.) converges on its movement towards Hutatma Chowk and further South and the Ballard Estate in Mody Back Reclamation, carry the lion's share of retailing. To the east and west of this core, retailing declines sharply, more so to the west of D.N. Road that constitutes the later additions to the Fort. Also, the commodity types vary between the commuter frontages on the main roads and feeder roads, the commodity trade that cannot stand the higher rent on the main shopping fronts of major roads retreating to the side roads.

It is also notable in Fort as a whole that retailing is essentially a ground floor activity. Practically, no retailing is found on the first or higher floors because of the onrushing character of the pedestrian commuter flows that have no time to spare to move vertically up in search of retail service. It is only on some by-lanes immediately adjoining D.N. Road that some specialised retailing is found on first floor in commodities like cameras, radios and videos that have an elite and searching clientele.

8.4 Internal Structure of the Retail Cluster of Mandvi-Bhuleshwar-Kalbadevi

The compiled areas under different urban land-use functions on both floors in case of Mandvi-Bhuleshwar-Kalbadevi are presented in Table 8.5. The areas under various tertiary and non-tertiary urban functions in the area for the two floors are shown in Maps 8.4 and 8.5.

8.4.1 *Land-Use Proportions on the Gross*

On the ground floor of the area as a whole, the predominant function is trading on the gross, but the retailing component, however, is not as significant as in the Fort area. Retailing occupies only 19.73 per cent of the total floor area, while the non-retailing trading function occupies a much larger share (nearly thrice) of 60.71 per cent; out of it, 18.37 per cent is occupied by wholesaling, 25.29 per cent is occupied by semi-wholesaling, 8.20 per cent by commission agents and 7.96 per cent by transport services, and a very small per centage of 0.89 per cent by godowns. Non-trading functions occupy only 15.73 per cent of the total area. Non-teritary functions in the form of religious and cultural areas occupy only 3.82 per cent of the ground floor area. Significantly, there is practically no residence or administrative service function on the ground floor.

As against this obtaining pattern on the ground floor, on the first floor, there is no retailing function at all. Non-retailing trading services occupy 37.74 per cent of the first floor area in this zone, of which wholesaling occupies 10.81 per cent semi-wholesaling 6.07 per cent

TABLE 8.5: Internal Land-use Structure of Bhuleshwar-Kalbadevi-Mandvi Zone

(All figures are in percentages of area under each function to the total area under all urban functions on each floor)

Functions/Services	Ground Floor %	First Floor %
1. Tertiary Services		
A. Trading		
(a) Retailing	19.73	—
(b) Non-Retailing		
(i) Wholesaling	18.37	10.81
(ii) Semi-wholesaling	25.29	6.07
(iii) Commission Agents	8.20	15.69
(iv) Transport Services	7.96	5.17
(v) Godowns	0.89	
Sub-Total (A)	60.71	37.74
B. Non-Trading		
(i) Banks and Financial Institutions	2.53	2.03
(ii) Commercial Establishments and Business Houses	13.20	28.67
(iii) Govt. Offices	—	—
Sub-Total (B)	15.73	30.70
Total Tertiary Services (A+B)	96.17	68.47
2. Non-Tertiary Services:		
(i) Residence	—	29.00
(ii) Religious and Cultural Centre	3.82	2.56
(iii) Educational Centres	—	—
(iv) Light Industry	—	—
(v) Others	—	—
Total Non-Tertiary Services	3.82	31.56
Grand Total (1 + 2)	100.00	100.00

Source : Data from Field Survey

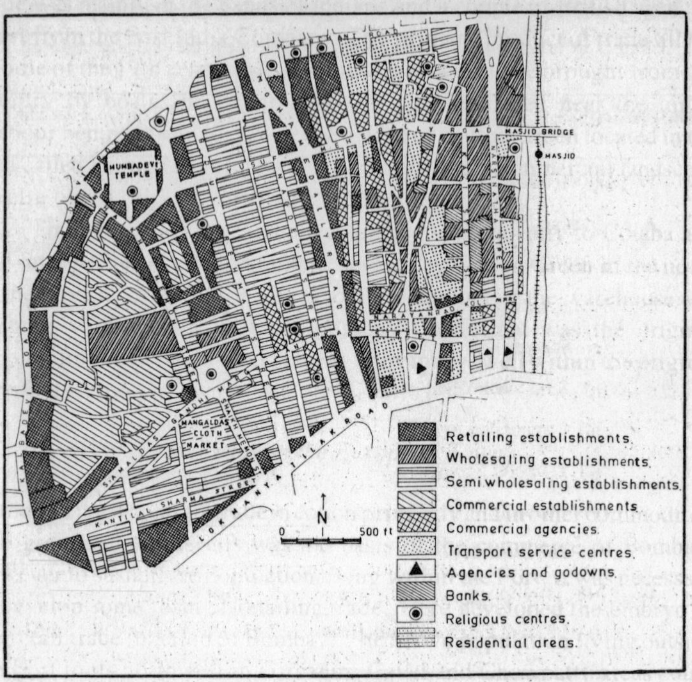

MAP 8.4: Land-Use Map of Mandvi Chakla and Market: (Ground Floor)

commission agents 15.69 per cent and transport services 5.17 per cent of the total area.

Non-trading tertiary services are more widespread on the first floor of the zone. Commercial establishments dealing with mail orders occupy 28.67 per cent of the total first floor area, and 2.03 per cent is under banks and financial institutions. The total area under tertiary services is 68.44 per cent.

Residence account for 29 per cent which is slightly more than the floor space occupied by commercial establishments. About 2.56 per cent of the first floor area is under religious and cultural centres.

In spite of a considerable share of space occupied by trading function on the first floor, retailing is totally absent, as in the Fort area. This zone has, however, a significant difference from the Fort area in the pattern of tertiary land-use, with a greater inter-mix of trading and non-trading functions, but is similar in that retailing is solely on ground floor, probably typifying the Indian urban context, in contrast to the western multi-storeyed retail complexes with dominance of departmental stoes and chain shops.

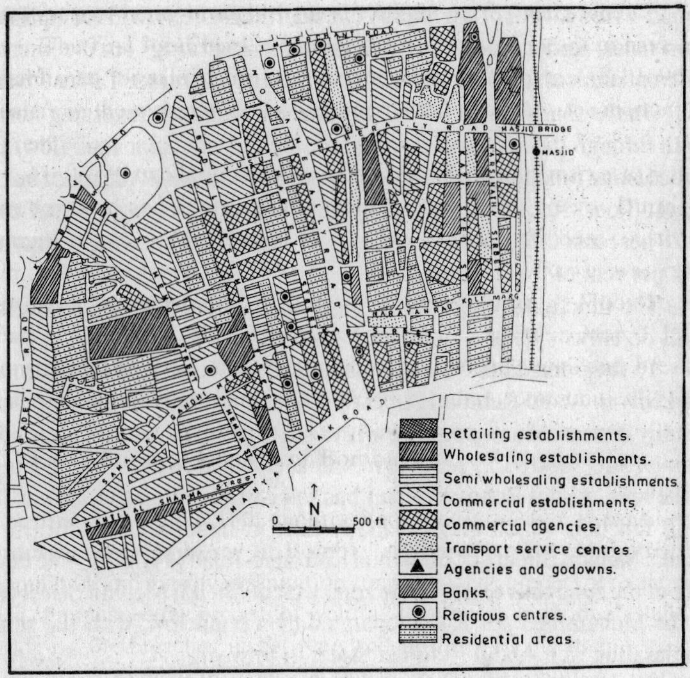

MAP 8.5: Land-Use Map of Mandvi Chakla and Market (First Floor)

8.4.2 *Pattern of Retailing Distribution within the Zone (Bhuleshwar, Kalbadevi and Mandvi)*

A road-wise analysis of the area by considering thirteen roads comprising main roads, feeder roads and by-lanes in the area brings out the internal differentiation of retailing and other land-use structure (see Tables 8.6 and 8.7). On the main road, that is Mohammed Ali Road traversing north-south through the area and carrying the bulk of vehicular road flow, the predeominat function is retailing, occuping 76 per cent of the space on the ground floor. Religious places (mainly mosques) occupy a significant 14.28 per cent of the floor space, Next to it, is semi-wholesaling occupying 9.52 per cent of the total area.

The functional character is somewhat different in the secondary roads on either side of Mohammed Ali road.

To the immediate west of Mohammed Ali Road, retailing declines markedly in Abdul Rehman Street (5.3%) but further west of it, retailing steadily increases in importance with a share of 46.44 per cent in Sheikh Memon Street and 86.29 per cent in Kalbadevi Road on the Western fringe of the area. Abdul Rehman Street has semi-wholesaling (84.2%) as the main function which also gradually falls in importance westward in Sheikh Memon Street (42.60%) and Kalbadevi Road (11.16%). The retail core of the zone thus exists in the zone west of Sheikh Memon Street and in the Mohammed Ali Road separated by a retail low, with the semi-wholesaling rich Abdul Rehman Street in between.

To the east of Mohammed Ali road, in Samuel Street, Kazi Syed Street and Narsinath Street, neither retailing nor semi-wholesaling is the main function. The predominant function is wholesaling. In Samuel Street, 78 per cent of the total area is occupied by wholesaling, in Kazi Syed Street 60 per cent and Narsinath Street 68 per cent. This wholesaling function is followed by transport services as its subsidiary in terms of floor space occupation. In Samuel Street, Kazi Syed Street and Narsinath Street, transport services occupy 22, 13.76 and 11.6 per cent respectively of the total floor space. Banks and financial institutions occupy a significant proportion of the area (17.43%) of Kazi Syed Street.

In the feeder roads to the west of Mohammed Ali Road (Nagdevi Street, Mangaldas Road, Kantilal Shah Marg), the main function is semi-wholesaling, and retailing follows as a poor second. For example, in Nagdevi Street, semi-wholesaling function occupies 72.16 per cent of the total area, while retailing has 15.46 per cent. On Mangaldas Road, semi-wholesaling accounts for 66.67 per cent and retailing for 33.34 per cent. On Kantilal Shah Marg, retailing function explains only 16.67 per cent

TABLE 8.6: Road-Wise Distribution of Land-use Functions within Bhuleshwar-Kalbadevi and Mandvi Area Zone-II - Ground Floor

(in percentage)

	Retailing	Wholesaling	Semi-Wholesaling	Commission Agent	Transport Agent	Godowns	Bank & Financial Institutions	Commercial Establishment & Business House	Govt. Office	Religious & Cultural Centre	Educational Institute	Recreation Centre	Light Industry	Residence	Others
I. Primary Road:															
Mohammed Ali Rd.	76.19	—	9.52	—	—	—	—	—	—	14.28	—	—	—	—	—
II. Secondary Roads:															
West of Mohammed Ali Rd.															
(a) Abdul Rehman St.	5.3	—	84.20	10.50	—	—	—	—	—	—	—	—	—	—	—
(b) Sheikh Memon St.	46.44	7.65	42.60	—	—	—	3.27	—	—	—	—	—	—	—	—
(c) Kalbadevi Rd.	86.29		11.16	—	—	—	2.53	—	—	—	—	—	—	—	—
III. Secondary Roads:															
East of M.Ali Rd.:															
(a) Samuel St.	—	77.77	—	—	22.22	—	—	—	—	—	—	—	—	—	—
(b) Kazi Syed St.	4.58	60.10	4.50	—	13.76	—	17.43	—	—	—	—	—	—	—	—
(c) Narsinath St.	—	68.10	5.80	10.10	11.60	—	1.40	—	—	—	—	—	—	—	—
IV. Feeder Road															
West of M.Ali Rd.:															
(a) Nagdevi St.	16.46	—	72.16	5.15	-	-	-	-	-	4.12	—	—	—	—	—
(b) Mangal Das	33.34	—	66.67	—	—	—	—	—	—	—	—	—	—	—	—
(c) Kantilal Shah Marg	16.66	—	83.30	—	—	—	—	—	—	—	—	—	—	—	—
V. Feeder Rd.															
East of M. Ali Rd.:															
(a) Bhandari St.	26.47	14.70	11.76	17.64	29.41	—	—	—	—	—	—	—	—	—	—
(b) Dhobi St.	11.11	33.33	22.22	33.33											
(c) Mirchi St.	—	—	90.90	—	—	—	9.09	—	—	—	—	—	—	—	—

TABLE 8.7: Road-Wise Distribution of Land-use Functions within Bhuleshwar-Kalbadevi and Mandvi Area Zone-II - First Floor

(In percentage)

	Retailing	Wholesaling	Semi-Wholesaling	Commission Agent	Transport Agent	Godowns	Bank & Financial Institutions	Commercial Establishment & Business House	Govt. Office	Religious & Cultural Centre	Educational Institute	Recreation Centre	Light Industry	Residence	Others
I. Main Road:															
Mohammed Ali Rd.	—	—	—	20.32	—	—	—	30.89	—	14.28	—	—	—	34.29	—
II. Trunk Roads :															
West of Mohammed Ali Rd.															
(a) Abdul Rehman St.	—	14.30	11.80	34.00	—	—	—	39.90	—	—	—	—	—	—	—
(b) Sheikh Memon St.	—	14.40	33.14	3.30	—	—	4.90	44.19	—	—	—	—	—	—	—
(c) Kalbadevi Rd.	—	16.76	—	—	—	—	2.99	80.23	—	—	—	—	—	—	—
III. Trunk Roads:															
East of M.Ali Rd.:															
(a) Sherif Deoli Rd.	—	—	—	—	—	—	—	—	—	—	—	—	—	—	—
(b) Samuel St.	—	—	—	22.22	11.11	—	—	33.33	—	—	—	—	—	33.33	—
(c) Kazi Syed St.	—	26.60	—	22.20	—	—	24.77	—	—	—	—	—	—	26.58	—
(d) Narsinath St.	—	46.70	—	18.10	7.80	—	6.00	—	—	—	—	—	—	21.50	—
IV. Lanes West of Mohammed Ali Rd.:															
(a) Nagdevi St.	—	—	—	37.63	—	—	—	—	—	4.30	—	—	—	58.06	
(b) Mangal Das	—	—	50.00	50.00	—	—	—	—	—	—	—	—	—	—	—
(c) Kantilal Shah Marg	—	33.33	—	58.33	—	—	—	—	—	—	—	—	—	—	8.33
V. Lanes East of Mohammed Ali Road:															
(a) Bhandari St.	—	—	—	—	29.41	—	—	—	—	—	—	—	—	—	57.14
(b) Dhobi St.	—	—	—	45.45	45.45	—	—	9.09	—	—	—	—	—	—	—
(c) Mirchi St.	—	8.69	—	39.13	17.39	—	34.78	—	—	—	—	—	—	—	—

as against 83.3 per cent of semi-wholesaling.

To the east of Mohammed Ali Road, the association of functions in feeder roads is quite different from that on the west side. Two important trading functions, retailing and wholesaling, are present in close association with semi-wholesaling, commission agents and transport service. In Bhandari Street, 26 per cent of the total area is occupies by retailing 14 per cent by wholesaling 12 per cent by semi-wholesaling 18 per cent by commission agents and 29 per cent by transport service. In case of Dhobi Street, wholesaling and commission agencies occupy 33.33 per cent area each and the rest of the area is divided between retailing and semi-wholesaling. But, in case of A.D. Gandhi Marg (Mirchi Street), the main function is semi-wholesaling. It is thus seen from east to west of this zone, on the ground floor, there is a gradual change of function from wholesaling to semi-wholesaling to retailing.

But on the first floor, there is a wider mixing of tertiary and non-tertiary functions. Except for religious establishments on Mohammed Ali Road, in all the north south trunk roads, west of Mohammed Ali Road, and in Mohammed Ali Road, there is no non-tertiary land-use at all on the first floor. Practically, the entire ground and first floor land-use function is commercial. Except in A. Rehman Street, in all of them retailing is predominant on ground floor, while semi-wholesaling (mail order), commission agents and commercial establishments practically account for the bulk of floor space use. On the other hand, in the trunk roads east of Mohammed Ali Road, residences, invariably of the shopkeepers trading in the area, account for a sizeable floor space-a fifth to a third, the rest is shared by non-retailing trade function. In the feeder roads, area under residence goes upto almost 50 per cent; banks and commission agents occupy almost the rest of the floor space.

8.5 Main Findings

A comparison of the retailing land-use characteristics between the two case study areas is rewarding in the findings.

(i) In the Fort area, the major land-use function of the ground floor space is retailing, while in Mandvi-Kalbadevi-Bhuleshwar, semi-wholesaling and retailing in that order are both significant and followed closely by wholesaling.

(ii) While in Fort area there exists a strong areal association between retailing, business houses and commercial establishments, in Mandvi-Kalbadevi-Bhuleshwar area, the different components of trade function, together with their aiding auxiliaries like commission agencies and road transport agents, form a mutual linkage, an integrated trade package, sharing the common advantages of the external, urban economies.

(iii) In both area, in spite of being the major areas of retailing clustering, retailing is entirely a ground floor activity. No retailing worth the name takes place on the first or higher floors, and this is possibly due to the lack of large-scale organisational enterprise in retailing in the form of departmental stores and chain shops and the almost exclusive small and medium individual private enterprises. The first floor space is occupied primarily by residence of shopkeepers and workers in the Fort area, while in Mandvi-Kalbadevi-Bhuleshwar, other trade functions including semi-wholesaling and commission agencies compete for the floor space.

(iv) Retailing in Fort has a historical association with the trading zone of the early colonial days; it has fossilised in the midst of a busy hive of administration, finance and commerce. On the other hand, the Mandvi-Kalbadevi-Bhuleshwar zone strongly reflects all the elements of a mature commercial landscape with strongly in-built urban and commercial linkages reflecting strongly in the intra-area land-use character.

REFERENCE

Murphy, R.E. (1974): *The American City, An Urban Geography*, Second Edition, McGraw Hill Book Co., pp. 345-376.

9

RETAILING LANDSCAPES IN INDIAN CITIES: An Overview

9.1 Utility of the Present Study

The systematic empirical analysis of the retailing landscape of Greater Bombay attempted in the preceding chapters reveals a number of interesting and useful findings that may be of assistance in broad generalisations concerning the evaluation of retailing structure and morphology in the Indian metropolitan cities. The findings, first of all, reflect the phases of evolutionary forms of spatial growth and multiplication in number, size and variety of retail areas in response to the temporal, demographic and functional growth, areal sprawls and suburban development along the arterial route ways, albeit, the spatial inertia that contributes to the continued survival and intensity of throb of the early retail core. There is thus witnessed a time-lag between the development of newer residential areas and the distinct emergence of a viable retail shopping area in the prevalent unplanned and haphazard rise and growth of new residential areas along the ever-widening urban fringes of the city. No doubt, local topographical and configuration constraints bring about eccentricities of evolution that are unique to individual cities.

The spatial order in the morphology and structure of retailing as well as the spatial processes that bring about this order also become evident. Arising out of the spatial order of retailing, a hierarchy evolves in terms of size, internal differentation and other characteristics.

9.2 Major Findings of the Study

The major findings of the present study can thus be summed up:

1. The embryo of the first retail shopping area developed within the European Colonial Fort area, on its northern side, within the ramparts, overlooking the native township across the Esplanade. It was an adjunct to the wholesaling export trade and port functions of the colonial port-city, essentially to serve the locally resident population which was largely European. A distinct commercial area, with commerce as its prime activity, emerged early in the 19th century in the Masjid-Mandvi area, just outside and north of the Fort, beyond the Esplanade, close to the docks and adjoining the native quarters of residence. It is this commercial complex that has grown in strength and intensity of function over the decades. In spite of the fact that the city has grown phenomenally in its population size and urban spread and this commercial complex has become an eccentric location in the south, close to the city centre, it still accounts for over half of the retailing activity of Greater Bombay, which has also increased manifold in volume and variety.
2. Further diffusion of retailing was initially mainly as probing linear intrusions along the main roads beyond the Fort and the commercial area of Mandvi-Masjid into the old residential cores. The development of a distinct retailing shopping area around Dadar rail-head was a sequel to the development of early suburbs in the north of the island city beyond the industrial fringe late in the first half of the 20th century. This response is possibly the first indicator of the impact of the friction of distance being felt on procurement of daily needs.
3. It is the process of retail suburbanisation along the eastern and central rail corridors in the post fifties of this century and the absorption of large areas of nearby Salsette into the urban fold, with the formation of Greater Bombay, that brought in its wake retail shopping zones along the commuting suburban rail-heads that are located increasingly further away from the commercial core of the city. However, a distinct time-lag is noticeable in the development of the retail shopping area, often two decades, that is, till a sizeable commuting population demanding retailing service is generated, in that suburban neighbourhood. By and large, areas away from the rail head have not developed shopping areas.

   Though the early growth of the colonial city of Bombay was heavily dependent on wholesaling trade and the port functions, its transformation into a premier industrial node of the country, especially of textiles, engineering, chemicals and petrochemicals, has however not reduced its commercial importance. The diversification and increasing intensity of other urban functions have only added to its commercial strength and brought about wide ramifications of linkages of an inherent commercial character that get reflected also in its retailing landscape.

5. With the alarming growth of population, especially since the forties of this century, mainly due to in-migration and the urban sprawl, the residential function of the suburbs has increased to a vast extent. As a result, more than half the population of Greater Bombay today lives in the suburbs. Simultaneously, the residential function of the city centre and the commercial core of A, B and C wards has declined, being pushed out by the increasing competition and pressure of high rental, non-residential urban functions.
6. Slightly more than a fifth of the work force of Greater Bombay is engaged in trade and commerce sector, and about 11 per cent in the retailing trade; this share is relatively large in comparison to most million cities of the country. Trade and commerce ranks first in 12 sections of Greater Bombay, mainly in B and C wards and adjoining parts of D ward. The highest concentration of retail index, density and intensity of retail employment are also noted in ward C and Fort North in ward A, while it is secondary only to wholesaling in the adjoining ward B. Retailing intensity and density are low in the industrial Central Bombay and the Suburbs.
7. The overwhelming importance and commercial significance of B and C wards and a part of A ward (Fort North) in retailing trade are also borne out by an investigation of the relative intensities of varied urban and commercial service functions and their mutual affinity and association as well as the Floor Space Index (FSI) values. A statistically significant positive correlation is noted at section level between retailing and wholesaling functions. Retailing also finds mutual association with high FSI Values. However, a significant negative finding, worthy of mention, is the weak positive relation between retailing and residential function which is obviously due to the infantile and insignificant development of retailing clusters in the suburbs that hold more than half the residential population, and the continued and continuing strength and intensity of retailing in the residentially declining CBD and commercial core that have an intense influx of daytime commuters.
8. Unlike the planned western cities wherein retailing is organised in central shopping areas, with large enterprise, departmental and chain stores in Greater Bombay, a sizeable quantum of retail trade, possibly a third, is in the informal sector associated with low-income population, and a large proportion is dispersed, individual, small enterprise corner shops. Retail shopping clusters are relatively slow to evolve till a sizeable consumer service population, above a

minimum threshold, is generated at spatially convergent location, like a suburban commuter rail-head. There exists a broad distance-decay from city-centre of clustered retailing in shopping areas. However, retail intensity develops areally a wave from, with the wave amplitude declining towards urban periphery. Relatively larger retail clusters occur at intervals, spaced with intervening retail poor areas.

9. Based on field-work and identification of retail shopping clusters all over Greater Bombay, 22 retail cluster zones have been identified. The emergent retailing landscape reveals its spatial organisation in four distinct spatial forms:

   (a) a predominantly commercial area near the city centre, wherein the area breathes commerce in its different commercial functions, such as wholesaling, retailing, warehousing, mail order functions, business function, banking and transport services, mutually reinforcing each other and in which retailing is a vital component of varying degrees of importance,
   (b) linear shopping fronts, especially on the ground floor, along the main arterial roads like the Colaba Causeway, Dadabhoi Naoroji Road, Kalbadevi Road, Jagannath Shankar Sheth Road, Bhadkamkar Marg and others with retail nodes in Chowks and Circles,
   (c) linear shopping roads, converging on commuting rail-heads in the suburbs, and
   (d) an amorphous, not too compact, shopping complex in some planned neighbourhoods.

10. Based on the areal density of retail shops in the retail clusters and the number of commodity functions incident from among 26 main commodity groups, as well as on the basis of a quantitative scoring based on a cumulative weightage score of retail commodity groups, centrality index or location co-efficient of retail function, an index of retail functional strength, and a correlation of retail functional index with the total floor area, it is established that a five-level retail spatial hierarchy is perceived in the retailing landscape of Greater Bombay, with Fort North and Bhuleshwar-Kalbadevi at the highest level, Mandvi-Chakala, Girgaum, Tardeo-Byculla and Dadar predominantly mid-level residential areas form the second level; Colaba to the south of the Fort, and some suburbs like Chembur, Bandra-Santacruz (West) and Goregaon-Malad (West) constitute the third level. The industrial Central Bombay, dock areas, as well

as most inner suburbs form the next level, while Borivli-Poisar-Dahisar (East) and Bandra (East)-Kalina are at the lowest level, due to late history of residential development.

11. While a significant positive association between population density and retail strength is noted, however, the relation between the hierarchic rank of retail clusters and the socio-economic status of the resident population is not so distinct partly because clearly identifiable, homogeneous socio-economic zones at the section level are not possible due to the free intermingling of the high, middle and low levels of residences and hutments all over the city and partly because most retail cluster service areas of the city, by virtue of the arterial nodal locations, attract sizeable clientele from beyond the areas as well.
12. Population densities in Greater Bombay have a distance decay from the city centre, which itself is a crater with depressed density values. Correspondingly, a distance-decay of retailing strength is noted, but without a central crater. In fact, the population density crater is a zone of highest retailing concentration; this fact is associated with its CBD and business functions and commuter inflows of service and employment seekers. While only 53 per cent of the clustered retailing intensity is explained by distance-decay, 74 per cent is explained by population densities. The distance-decay broadly confirms the continued strong presence and survival of a colonial core near the city centre and the more recent emergence of retailing amenities, yet in a youthful form, in response to the pressure of growing residential population and newer waves of suburbanisation.
13. A detailed field mapping and analysis of six main retail cluster zones belonging to the different hierarchic levels except the lowest order reveals an age gradient in retailing from the city centre outward towards the urban fringe. Though essentially an economic activity, mostly of individual small-scale enterprise, retailing in Bombay shows strong linguistic, religious and even caste affinities, often with distinct commodity specialisation. There also exists a shop size-commodity relation with jewellery, textiles and machine parts as the large ones with larger money investment. Serviced clientele also varies with commodity. So also there exists a broad relation between the socio-economic class level of residential areas and shopkeepers' residence. Individual retail shopping clusters reveal characteristic associations of social groups, commodity functions and consumer types.
14. Lastly, an intensive plot-to-plot field analysis of retail shopping areas in Fort and Kalbadevi-Bhuleshwar-Mandvi reveals an internal

differentiation of retail shopping land-use. Micro-level variations of site location reveal contrasts of development between the main and feder roads and the by-lanes as well as between the corridors of pedestrian commuter flows and those less frequented by commuters. In both the areas, retailing is essentially a ground floor activity and is not developed vertically, quite unlike the western cities, though Bombay is the most westernised cosmopolitan city of India. The internal land-use differentiation and mutual associations in the commercial core bring out the intimate interaction and interrelation between the different components of trading and non-trading activities, including non-tangible services like hotelling, beauty parlours and the like.

9.3 A Schematic Generalisation of Retailing Landscape in Greater Bombay

Deriving from the study a schematic generalisation of the clustered retailing landscape in relation to the city's urban functional morphology and spatial growth can be evaluated, as shown in Fig 9.1. The urban ecological model of Greater Bombay reveals a concentric ring pattern developed as an areal sector, thanks to its insular location and the eccentric siting of the CBD in the Fort, near the south end. The arterial road-cum-rail corridors have thus a north-south lineation, along with all subsequent growth has taken place.

The city centre, fringed by the old residential core, has an external ring of industrial zone, succeeded by three peripheries of waves of suburbanisation—early, mid and late 20th century sprawls. The commercial core, close to the rail terminal and the docks on the eastern front, adjoining the old native and European urban cores. It contains even today over half the retailing activity. The fist outward growth of retailing is in linear road corridors in the city centre and its adjoining old residential cores and extensions. A significant outward shift of retailing is the early 20th century retail cluster at the joint rail-head of Dadar to develop a miniature replica of the older retail complex of South Bombay. This is in response to significant residential development in the north of the city. Further growth of suburbs in successive waves along the main rail corridors brought rail-head retail areas in some, better developed suburbs, such retail areas being spaced at 3 to 51 km distances along the railway. Interstitial growth is continuing, but the heavy concentration of retailing still within the old city core and a belated and lagged development of shopping areas in the suburbs still in its infancy are evident.

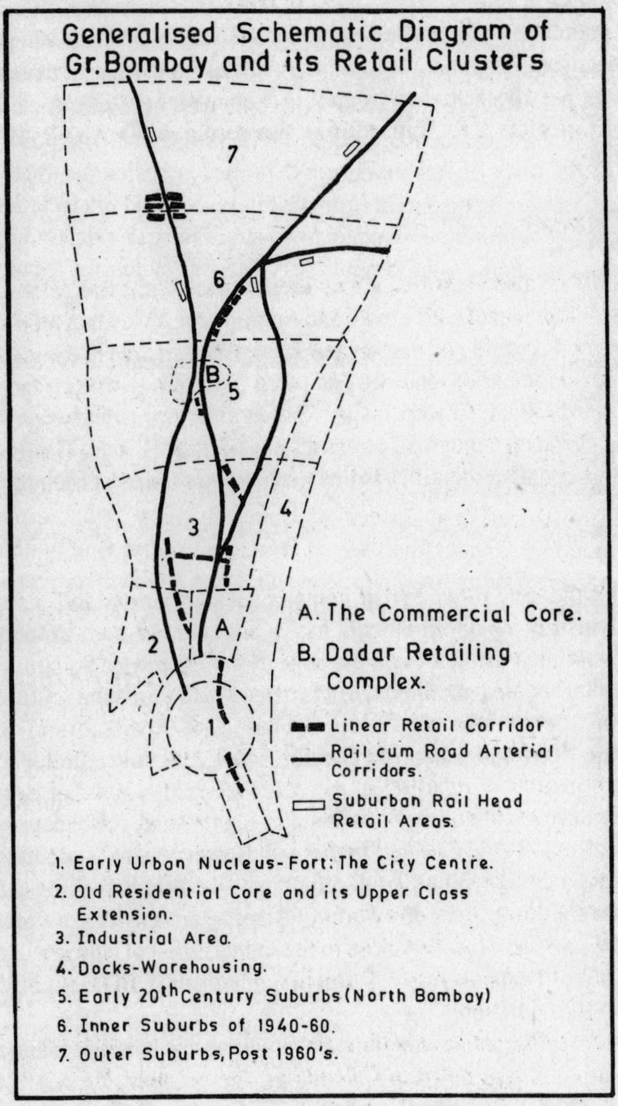

FIG. 9.1 : A Schematic Generalisation of Retail Landscape in Greater Bombay

9.4 Retailing Landscape in Some Major Indian Cities: A Comparison

9.4.1 Studies of retailing landscape in other Indian metropolitan cities with a population of a million or more are scanty and not adequately comprehensive for generalisation of retailing landscape. Calcutta and Bangalore have been studied (Map 9.1) to an extent in this regard. In his study of Calcutta, A.K. Dutt, (1967)[1] has recognised a four-level retail hierarchy.

9.4.2 *Bangalore*

Sastry (1981)[2] also identified a four-level hierarchy for Bangalore, with a twin development of native town and cantonment. Two main retail areas are identified: the old city market and Russel market area of cantonment. Linear development of retailing has taken place along major roads like Kempegowda Road, Chickpet, City Market area, Narasimaharaja Road, M.G. Road and on major road crossings like Majestic Circle. The five-tier hierarchy of Bombay identified in this study is thus to an extent comparable.

9.4.3 *Calcutta*

Calcutta,[3] the only other city of comparable dimensions and a similar colonial histroy of development, has a commercial core around the Burrabazaar area (Map 9.1) on the edge of the Dalhousie Square. From the Burrabazaar area, retailing clusters extend both north and south along the main arteries like the Netaji Subhas Bose Road, Strand Road, Brabourne Road and Mahatma Gandhi Road. It is quite similar to the linear protrusions of retailing in the older residential parts of Bombay along major roads like Jagannath Shankar Sheth Road, Kalbadevi Road, D.N. Road and Shahid Bhagat Singh Road. However, since Calcutta does not possess a similar suburban development like Bombay, the rail-head commuter shopping areas of suburbs of Bombay are missing in Calcutta. The retail core has a location close to the rail terminal of Howrah like the market area of Bombay near V.T. But in Calcutta, the CBD is slightly free from retailing function.

Internal differentiation of the main retailing areas based on commodity specialisation is also noted in Calcutta as, for example, Ezra Street for electrical goods, Barabourne Road for crockery, College Street for printing and publications.

9.4.4 *Madras*

The third of the old colonial port cities of East India Company with a

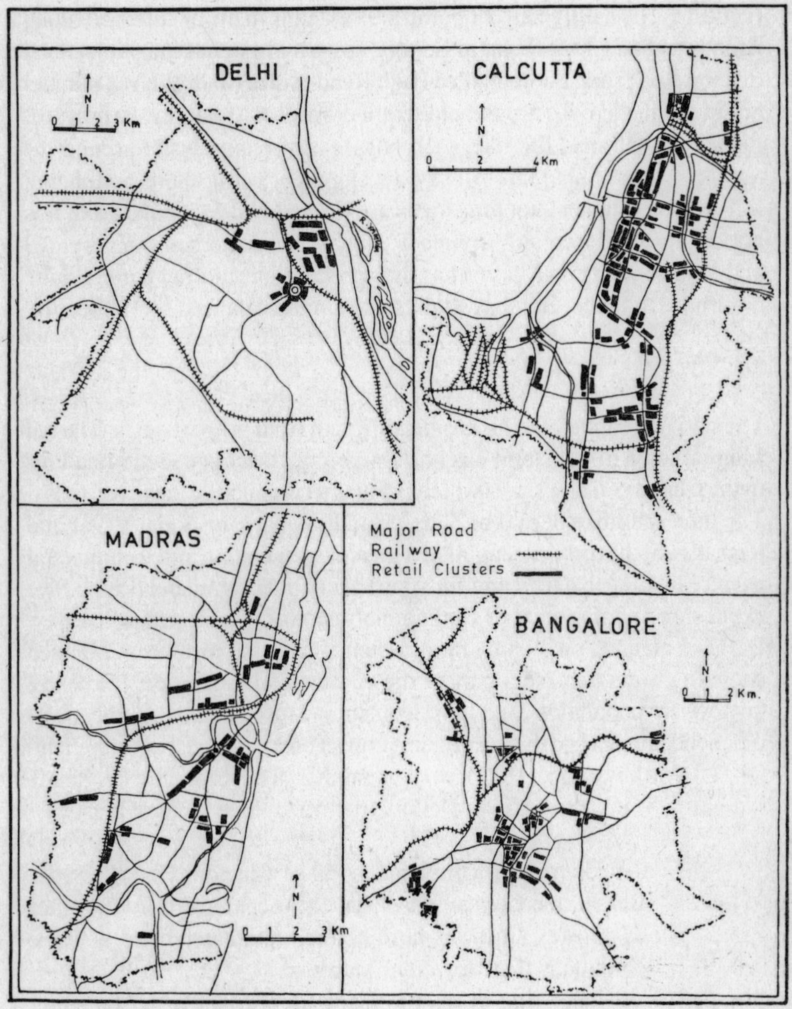

MAP 9.1 Retail Land-use of other Cities

comparable colonial history, Madras[4] has, however, a more spreadout development of retailing shopping areas. The main wholesale complex centres around park town around Bandar Street with an intermixture of retailing. The main retail shopping area extends in linear corridors along Broadway and China Bazar in George Town and extends into Anna Salai (Mount Road) and Poonamallee High Road. Since Madras has a number of Urban nuclie built around old temple towns of Mylapore, Triplicane, Purasawakkam and the like, each has its own retail cluster around the temple streets. Suburban railway development being more peripheral, rail-head commuter shopping areas are limited in development and it is noticeably developed only around T. Nagar. The pattern of urban sprawl and road corridors in Madras has thus produced a retailing pattern quite dissimilar to that of Bombay. It presents a more balanced spatial scatter.

9.4.5 *Delhi*

The national capital, Delhi[5] is quite different from the port cities. The old Mughal town of old Delhi has its commercial area outside the Red Fort in the Chandni-Chowk area which is a major commercial area even today. The then suburban bazaar of Subzi-Mandi, Paharganj, Sadar Bazar and Kishangunj have been engulfed by later residential and commercial development so that they now form part of an inner commercial zone. The expansion and dispersal of commercial use was accelerated when New Delhi, a planned city, came into existence with the result that planned shopping areas became a part of the commercial landscape. Explosive growth in population and physical spread of the city in the post-independence era had its concomitant impact on commercial development. Dr. R. Parthasarathi (1971)[6] in her study, "The Commercial Retail Shopping Structure of Urban Delhi", identified three types of clusters:

(i) To commercial core, consisting of Chandni Chowk, Sadar Bazaar and Connaught Place, each separated from the other by railway lines or yards, Here, commercial use is more ubiquitous than any other land-use;

(ii) An inner ring or mantle made up of Mandvi, Kishangunj, Paharganj and Karol Bagh. Commercial use here is as ubiquitous as residential use;

(iii) The peripheral zone containing inner and outer rings, where commercial use is restricted to smaller well-defined, planned clusters, aligned to residential areas and arterial roads.

Retailing is the most widespread of commercial activities; it also conforms to the general tendency of larger concentration in the central area of the city. The major retail centres are Chandni Chowk, Connaught Place and the more recently developed Ajmal Khan Road zone in Karol Bagh. Chandni Chowk offers a rich variety of traditional goods; Connaught Place attracts a sophisticated clientele from all over Delhi, while Ajmal Khan Road specialises in textiles and catering to middle class population.

9.4.6 *Pune*

Pune differs from other million cities in that it has a linear corridor development of retailing on its fringe, especially Bombay-Poona Road, apart from the old retail core in Budhwar Chowk in the old city, with an elongated linear development along Lakshmi Road.

9.4.7 *Hyderabad-Secunderabad*

In case of Hyderabad-Secunderabad,[8] retailing is principally a string development. It strengthens at points of traffic intersection. The twin city has a number of city centres. Each CBD has a major shopping centre for its area with specialities of its own which makes it also central to the other city areas. For example, South has specially developed traditional items of bridal wear and jewellery. For engineering goods and automobiles parts, the CBD of Secunderabad serves the entire city.

9.4.8 *Ahmedabad*

In Ahmedabad,[9] the retail core lies within the old fortified town around Manek Chowk and Raton Pol area. An elite shopping spine has developed along Ashram Road to serve the newer residential areas outside the old city and across the Sabarmati.

9.5 Common Elements of Retailing Patterns in Indian Citiies

The above review of the retailing commercial landscape in some of the metropolitan cities of India indicates that these cities have at best a limited commonality, possibly because only a few of them have risen through a colonial commercial port tradition, and many have varied evolutionary histories. The common features are:

(a) the pressure of a distinct informal retailing sector, accounting for a sizeable servicing of the ubiquitous needs;
(b) a retail core, in or near the original city centre, often associated with other trade functions like wholesaling and warehousing

but at times with a weak development of wholesaling;

(c) ribbon development of shopping areas along a few main thoroughfares of traffic floors, mixed with residential and other urban functions; and

(d) a wide dispersal of corner shops catering to daily needs in all niches of the city, more so in residential areas. Suburban shopping complexes are not necessarily developed in all cases

A significatn departure is the planned input of centralised markets and shopping areas in Chowks in planned cities like New Delhi and Chandigarh, but these are the exceptions rather than the rule. With a long evolutionary history over centuries and a haphazard growth, spatial sprawl and agglomeration, under pressure of growth of urban population and functions, most retailing development is unplanned, sporadic and spontaneous to meet a growing need for such retail services, as distance friction begins to be felt in terms of time and cost, in reaching the main retail core or its extensions. In such a situation, emergence of retail rich and retail poor areas in the city is inevitable, adding consierably to urban problems of traffic congestion, imbalances of land-use and rental values.

9.6 Retailing Studies Needed

In the context of these cities, it is not possible to start on a clean slate to generate planned centralised retail amenities—nor is it feasible and desirable—but it is certainly possible to identify areas of priority development inputs for retailing amenities in the context of the flux of evolutionary elements present, and take advantage of such sites for reducing the imbalances of growth. The present study of the retailing landscape of Bombay provides some deep insights in this regard in the complex urban morphology of the commercial metropolis which should prove meaningful and useful for planners not only of this city, but also of other cities. What, however, would be warranted, is more detailed and systematic comprehensive studies of the commercial landscapes of all Indian cities by geographers and other social scientists before viable common urban policies of development can be outlined. It is earnestly felt that the present study of retailing complexes in Greater Bombay is a step in this right direction.

REFERENCES

1. Dutt, A.K. (1967): "Shopping in Calcutta", *Town Planning Review*, Vol. 37, pp. 207-216.

2. Sastry, G.S. (1981): "Retail Trade Structure of an Indian Metropolis, Bangalore", *Annals of the National Association of Geographers, India,* Vol. 1 No. 2, pp. 57-73.
3. Sinha, P. (1978): *Calcutta in Urban History.* Firma K.L.M. Private Ltd., Calcutta.
4. Madras City, "A Planning Study, 1960", *Urban and Rural Planning Thought,* Vol. 3 pp. 54-98.
5. Sundaram, K.V. (1978): "Delhi the National Capital", R. P. Misra (Ed.), *Million Cities of India,* Vikas Publishing House, New Delhi, pp. 104.154
6. Parthasarathi, R. (1971): "The Commercial Retail Shopping Structure of Urban Delhi", Ph.D. Thesis, Department of Geography, University of Delhi.
7. Joshi, R.V. (1952): "Urban Structure in Western India, Poona: A Sample Study", *Geographical Review of India,* 14th March, pp. 7-19.
8. Manzoor Alam (1963): "Commercial Structure of the Cities of Hyderabad-Secunderabad", *Journal of the Osmania University,* Vol. 1, p. 13.
9. Jain, K.C. (1978): "Ahmedabad—An Expanding Metropolis", R. P. Misra (Ed.) *Million Cities of India,* Vikas Publishing House, New Delhi, pp. 215-245.

APPENDIX 1

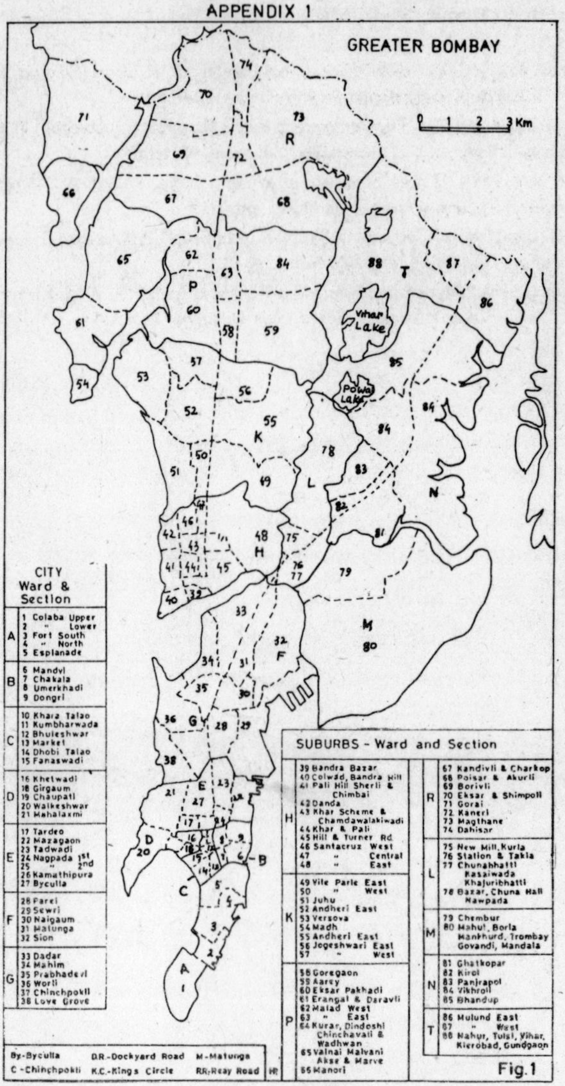

Fig.1

APPENDIX II

DETAILED COMMODITY CLASSIFICATION USED FOR ANALYSIS

The sample shops surveyed have been grouped into 26 broad commodity types based on detailed list of commodites dealt in, as given below:

1. Electrical and Related Commodities

- (i) Electrical Machineries and Fixtures.
- (ii) Electrical Appliances.
- (iii) Air Conditioners, Microprocessors.
- (iv) Stereo-Cassettes, TV and other Electronic Equipments and Appliances.
- (v) Radios and Transistors.
- (vi) Refrigerators, Washing Machines.
- (vii) Musical Instruments.
- (viii) Scientific Instruments.
- (ix) Pump Sets.

2. Grocery-Provisions

- (i) Grocery provisions of various types.
- (ii) Edible Oils.
- (iii) Spices, Jaggery, Sugar.
- (iv) Ration shop, Grainshop etc.

3. Textiles

- (i) Suiting and Shirting (Mill Show Room).
- (ii) Other dealers in Suiting and Shirting.

(iii) Silk Saree (Wedding Dress).

(iv) Other silk and Cotton Saree.

(v) All sorts of synthetic garment material.

(vi) Textiles in general.

(vii) Kiddies' Wear Centre.

(viii) Woollen Textiles.

(ix) Imported and Superior Garments.

(x) Textiles and Departmental Stores.

(xi) Jeans and Readymade Garments.

(xii) Hosiery and Knit Wear.

(xiii) Bed Sheets, Bed Covers, Pillow Cases etc.

(xiv) Carpets, Floor Mats etc.

4. Sweetmeat

(i) Brijwasi, Thakkar and Other Chain Stores of Sweetmeat.

(ii) Other Sweetmeat Stores.

(iii) Milk and Dairy products.

5. Footwear and Leather Products

(i) Dealers in all types of Footwear (Bata, Dadwood etc.)

(ii) Other Footwear Centre.

(iii) Leather Goods, Leather Wears, Suitcases, Ladies Handbags, Briefcases.

(iv) Seat Cover, Rexine Products.

(v) Foam Products, Upholstry, Coir Products.

6. Machinery, Spare Parts and Hardwares

(i) Engineering and Machinery items.

(ii) Automobiles (exclusive Show Room).

(iii) Other Automobile Spare Parts.

(iv) Bicycles.

(v) Three-wheelers, Scooters and Mopeds.

(vi) Asbestos and Tin Products.

(vii) Textile Machineries.

(viii) Hardwares, Sanitary-wares, Pipes etc.

7. Utensils

(i) Household Utensils.

(ii) Brass, Bronze and Copper Products.

(iii) All sorts of Stainless Steel Products.

8. Tobacco and Other Products

(i) Tobacco.

(ii) Pan, Bidi, Cigarette, Bread, Eggs etc.

9. Variety Goods

(i) Cosmetics.

(ii) Variety goods and Fancy goods of various types.

(iii) Novelties etc.

(iv) Antiques.

(v) Lampsheds.

10. Medicine and Drugs

(i) Chemists, Druggists.

(ii) Surgical Instruments.

(iii) Optical Stores.

11. Paper, Printing and Stationery

(i) Paper (for writing).

(ii) Paper Board.

(iii) Stationery and Printing Materials.

(iv) Packing Materials.

(v) Packing Cases.

(vi) Paper Bags.

(vii) Cardboard Boxes.

(viii) Covers and Envelopes etc.

12. Jewellery

(i) Diamond, Precious Stones and Jewellery.

(ii) Gold and Silver Jewellery.

(iii) Imitation Jewellery.

13. Sports Goods and Toys

(i) Exclusive dealers in Sports goods.

(ii) Other Sports goods shops.

(iii) Play Materials (exclusive dealers).

(iv) Toys and Play materials.

(v) Plastic goods including household items.

14. Crockery and Porcelain

(i) Crockery and Porcelain of all types.

15. Cold Storage, Dry Fruits and General Stores

(i) Dry Fruits.

(ii) Cold Storage (Mafco, Shalimar).

(iii) Sundry Food (Farsan, Mixtures, Biscuits etc.)

(iv) General Stores.

16. Vegetables and Perishables

(i) Vegetables of all varieties.

(ii) Fruit Stalls located at important street corners.

(iii) Dealers in all varieties of Flowers.

(iv) Flourists (specialised).

17. Sewing and Embroidery Materials

(i) Cutpieces.

(ii) Embroidery Materials.

(iii) Lace, Saree falls etc.

18. Watch, Clock, Pen and Gift Centres

(i) Watches and Clocks of all types.

(ii) Pen shops.

(iii) Gift Centres etc.

19. Books and Periodicals

(i) Books and Periodicals of all types.

20. Photography and Xeroxing

(i) Photo frame goods.

(ii) Camera, Films and Photographic Goods.

(iii) Advanced Photographic Equipments.

(iv) Xeroxing (Photocopying) and Laminations.

21. Paints and Chemicals

(i) Paints.

(ii) Chemicals

(iii) Cement and Building Materials.

(iv) Distilled Water.

(v) Pharmaceutical Goods.

(vi) Pesticides and Dyes.

22. Timber

(i) Timber.

(ii) Firewood.

(iii) Plywood.

(iv) Sunmica, Formica.

(v) Cane Products.

23. Furniture

(i) Furniture and Fixtures.

(ii) Steel Furnitures.

24. Bakery, Confectioneries etc.

(i) Bakery Products (Ordinary).

(ii) Specialised Cake Shops.

25. Perfumes

(i) Perfumes and "Aatars" (specially for Muslims).

26. Jari, Glass and Miscellaneous

(i) Glass and Mirror.

(ii) Lock and Key.

(iii) Ropes and Cords.

(iv) Rubber Stamp.

(v) Name Plates.

(vi) Jari and other Dress Materials.

(vii) Tin and Steel Boxes and Trunks.

APPENDIX III

QUESTIONNAIRE

1. Name of Zone : ______________________

2. Commodity/Commodities handled by the Shop (write Code No.): ______________________

3. How Many years ago it was established: ____________

(a) Less than 5 ( ) (b) Year of establishment: _____

5 - 10 ( )

11 - 25 ( )

26 - 40 ( )

41 - 55 ( )

56 - 70 ( )

71 - 90 ( )

90 ( )

4. What was the previous use of this premise?
   Residential/commercial

5. If it was residential, what was your status there:
   Owner ( ) Tenant ( ) I was not there ( )

6. At present what is your status:
   Owner ( ) Partner ( ) Employee ( )

7. If you are the owner from the beginning, why have you started this business:
   (i) Because it was family business ( )
   (ii) Because it is highly profitable ( )
   (iii) If other reason, specify: ______________________
   ____

8. If you are the Employee, where does the owner stay?
   Socio-economic status zone Name of the place:
   I
   II
   III
   IV
   V

9. Is it the main shop or a branch:

Main Shop ( )

Branch ( )

Only shop ( )

10 (a). If it is the main shop, where is the branch located?

Retail hierarchy zone	Name of the area
I	
II	
III	
IV	
V	

10 (b). If it is the branch, where is the main shop located?

Retail hierarchy zone	Name of the area:
I	
II	
III	
IV	
V	

11. What is your religion?

Hindu/Muslim/Christian/Parsee/Jain/Sikh.

12. What is your mother tongue?

Gujarati/Marathi/Sindhi/Parsee/Punjabi/Tamil/Malayalam/Kannada/Telugu/Hindi/Urdu/Rajasthani.

13. How many persons are working in the Shop?

10 ( )

6-10 ( )

3-6 ( )

1-3 ( )

Self-employed ( )

14. How many are employees?

15. Are these people working throughout the year or in some particular season?

(a) Throughout the year (b) For some particular season.

16. If (b) is "Yes", then in which season does the number increase?

Month	*Season*	*Social importance*
May-August	Monsoon	
Sept-December	Autumn/Winter	Ganpati, Diwali, Id, Christmas.
Jan-April	Late Winter/ Spring	Holi, New Year, Parsee New Year.

17. From where do you procure your commodities?

From the City	*Outside the City*
Warehouses/Godown/Factory	

18. What are the average number of customers visiting the shop per day?

25 ( )
25-50 ( )
50-100 ( )
100 ( )

19. During which part of the day, does the shop receive maximum customers?

10.00 a.m. to 12.00 noon ( )
12.00 noon to 3.00 p.m. ( )
3.00 p.m. to 5.00 p.m. ( )
5.00 p.m. to 8.00 p.m. ( )

BIBLIOGRAPHY

Books

Alam, S.M. (1965): *Hyderabad-Secunderabad : A Study in Urban Geography,* Osmania University Press, Hyderabad.

Alam, S.M. et al. (1972): *Metropolitan Hyderabad and Its Region,* Asia Publishing House, Bombay.

Anderson, P. (1856): *The English in Western India,* London.

Applebaum, W. (1954): "Marketing Geography", Preston E. James, and Clarence F. Jones (ed.), *Inventory and Prospect,* American Geography, University Press, Syracuse, N.Y.

Arunachalam, B. (1978): "Bombay—An Exploding Metropolis", in Misra, R.P. (Ed.) *Million Cities of India,* New Delhi.

Arunachalam, B. (1981): "Bombay", Michael Pacione (Ed.) *Problems and Planning in Third World Cities,* New Delhi.

Belcher, M. (1892): Statistich Beschreibung Der Studt, Frankfurt Am Main Und Threr Bevolekerung, Frankfurt Am Main.

Berry, B.J.L. (1962): "The Commercial Structure of American Cities: A Review", Community Renewal Programme, Chicago.

Berry, B.J.L. (1963): "Commercial Structure and Commercial Blight", Research Publication No. 85, University of Chicago, Chicago.

Berry, B.J.L. (1967): *Geography of Market Centre and Retail Distribution,* Prentice Hall, Englewood Cliffs.

Bhat, L.S. and other (1974): "Central Place Hierarchy for a Developing Agricultural Region: A Case Study of Karnal Area", Report of a Study sponsored by I.C.S.S.R., New Delhi.

Bhat, L.S. (1976): *Micro-level Planning: A Case Study of Karnal Area,* Haryana, India, K.B. Publisher, New Delhi.

Boekhalt, J.M.C. and Vanderveer, P. (1986): "Retail System of Vegetable in Mysore City, India" P.D. Mahadev (ed.) *Contribution to Indian Geography VII, Urban Geography,* New Delhi, pp. 110-131.

Bogue, D.J. (1949): The Structure of Metropolitan Community: A Study in Dominance and Sub-dominance, Ann Arbor, University of Michigan.

Brahme, Sulbha (1977): "The Role of Bombay in the Economic

Development of Maharashtra". A.G. Noble and A.K. Dutt (Ed.) *Indian Urbanisation and Planning.* Tata-McGraw Hill, New Delhi.

Carter, H. (1972): *The Study of Urban Geography,* Edward Arnold, London.

Christeller, W. (1933) *Die Zeutralen Orte in Suddentschland (Jena)* Translated by W.C. Baskine, 1966—Central Places in Southern Germany, Englewood Cliffs, N.J.

Davies, R.L. (1976): *Marketing Geography with Special Reference to Retailing,* Cambridge.

Didee, J. (1984): *Central Places of Western India,* Pune.

Deshpande, C.D. (1973): "Trends in Cities and Towns of Western India", Alam-Pakshishevsky (Ed.): *Urbanisation in Developing Countries.*

Desai, A. (1985): *Spatial Aspects of Settlement Pattern,* Concept Publishing Co., New Delhi.

Douglas, J. (1893): *Bombay and Western India,* Vol. I, London.

Edwards, S.M. (1902): *The Rise of Bombay—A Retrospect*, Times of India Press, Bombay.

Elwood (1830): *Narrative of a Overland Journey to India,* Vol. 2, London.

Falkland (1930): "Chow-chow", edited with Introduction and Notes by Rawlinson, Vol. 9, H.G., London.

Forbes, J. (1834): *Oriental Memoire,* Vol. 1, London.

Garner, B.J. (1966): *The Internal Structure of Retail Nucleations,* North Western University, Evanston.

Garrison et al. (1959) Empirical Verification of Concept of Spatial Structure, Studies of Highway Development and Geographical Changes, University of Washington Press, Seattle.

Graham, M. (1812): *Journal of Residence in India,* Edinburg,

Guha, B. (1964): *Morphology of Burdwan,* The Earth.

Hagget, P. (1972): *Geography: A Modern Synthesis,* Harper and Row Publishers, New York.

Hoover, E.N. (1948): *The Location of Economic Activity,* New York.

Hoyt, H (1939): "The Structure and Growth of Residential Neighbourhood in American Cities", in *Reading in Urban Geography,* Mayer H.M. & Kohn C.F. (Eds.) University of Chicago (1967).

Huff, D.L. (1962): *Determination of Intra-urban Retail Trade Area,* University of California, Los Angeles.

Isard, W. (1956): *Location and Space Economy,* Cambridge.

Johnston, R.J. (1978): "Residential Area Characteristics: Research Method for Identifying Urban Sub-area—Social Area Analysis and Factorial Ecology", Herbert, D.T. and Johnston, R.J. (Ed.) *Social Area in Cities, Process Pattern and Problem,* John Wiley and Sons,

Chichester.

Kar, N.R. (1962): Growth Distribution and Dynamics of the Population Load in Calcutta, (Mimeo).

Kevin, L. (1960): *Image of the City,* Cambridge, Mass.

King, M. & Cole, J.P. (1968): *Quantitative Geography,* John Wiley and Sons Ltd., London.

Losch, A. (1954): *The Economic of Location,* English edition New Haven, translated by William H. Waglam and Wolfgang F. Stopli, New Haven.

Mahadev, P.D. (1975): *People, Space and Economy of an Indian City,* Mysore.

Mandal, R.B. (1989): *Statistics for Geographers and Social Scientists,* Concept Publishing Co., New Delhi.

Mayer, H.M. and Kohn, C.F. (1967): *Reading in Urban Geography,* Central Book Depot, Allahabad.

McGee, T.G. (1971): *The Urbanisation Process in the Third World,* Bell and Sons, London.

McGee, T.G. and Yeung, M.M. (1969): *Hawkers in South East Asian Cities: Planning for the Bazaar Economy,* Ottawa.

Murphy, R.E. (1974): *The American City: An Urban Geography,* 2nd ed., McGraw Hill Book Company, New York.

Misra, H.N. (1977): "A Model of Economic Base and its Application to the Towns of Uttar Pradesh", P.D. Mahadev (ed.), Contribution to Indian Geography, *Urban Geography,* Vol. VII.

Postan, M. (1839): *Western India in 1838,* Vol. I, London.

Prakasa Rao, V.L.S. and Ramchandran, R. (1975): "Socio-Economic Survey of Vijayawada City, Mimeographed, ICSSR, New Delhi.

Prakasa Rao, V.L.S. and Tewari, V.K. (1979): *The Structure of an Indian Metropolis Bangalore,* Allied Publication, New Delhi.

Prakasa Rao, V.L.S. (1983): *Urbanisation in India: Spatial Dimension,* Concept Publishing Company, New Delhi.

Ralph, W.P. (1960): *The Techniques of Urban Economic Analysis,* Chandler, Davis Publishing Co., West Trentor.

Ratcliff, R.U. (1949): *Urban Land Economics,* McGraw Hill Book Company, New York.

Raza, Moonis and Habib, A. (1976): "Characteristics of Colonial Urbanisation—A Case Study of the Satellite Primacy of Calcutta (1850-1921) in Alam Pokskishevsky (ed.) *Urbanisation in Developing Countries.*

Rolph, I.K. (1929): "The Location Structure of Retail Trade", Domestic Comerce Series No. 8, United States Bureau of Foreign and Domestic Commerce.

Scott, P. (1970): *Geography and Retailing,* Hutchison, London.

Sherrat, G.G. (1960): "A Model for General Urban Growth", in Churchman, O.W. and Verhulst (Ed.), *Management Sciences, Model and Techniques,* Pergamon Press, New York.

Singh, R. L. (1955): *Benaras—A Study in Urban Geography,* Nand Kishore & Bros., Benaras.

Smailes, A.E. (1969): "The Indian City Structure", in Singh, R.L. (Ed.), Urban Geography in Developing Countries, *Proceeding* of I.G.U. Symposium No. 15, Varanasi.

Smith, D.M. (1973): *Geography of Social Well-being in the United States,* New York.

Spigel, R.M. (1961): *Theory and Problem of Statistics,* McGraw Hill Book Company, New York.

2. List of Papers—Published and Unpublished

Abu-Loghod, J.L. (1969): "Testing the Theory of Social Area Analysis: A Case of Cairo, Egypt", *American Sociological Review,* Vol. 34, No. 2, pp. 198-212.

Abiodun, J. (1947): "Urban Hierarchy of a Developing Country", *Economic Geography,* Vol. 3, pp. 347-367.

Ajo, R. (1965): "On the Structure of Population Density in London, Field Acta Geographic, Vol. 18, Cited by Prakasa Rao (1983): *Urbanization in India,* Concept Publishing Co., New Delhi. p. 99.

Andrews, R.B. (1953): "Mechanics of the Urban Economic Base", *Land Economics,* Vol. 29, pp. 161-167.

Applebaum, W. and Cohen, S.B. (1961): "The Dynamics of Store trading Area and Market Equilibrium", *Annals of the Association of American Geographers,* Vol. 51, pp. 83-106.

Aurousseau, M. (1921): "The Distribution of Population: A Constructive Problem, *Geographical Review,* pp. 563-592.

Berry, B.J.L. (1961) "A Method of Deriving Multifactor Uniform Region", *Prezeglad Geograficzny,* Vol. 33, pp. 263-282

Berry, B.J.L., Simmon, J.W. and Tennants, R.J. (1963): "Urban Population Densities: Structure and Change", *Geographical Review,* Vol. 53, No. 2, pp. 389-405.

Berry, B.J.L. and Rees, P.H. (1969): "The Factorial Ecology of Calcutta", *American Journal of Sociology,* Vol. 74, pp, 445-491.

Berry, B.J.L. and Garrison, W.L. (1958): "Functional Bases of Central Place Hierarchy", *Economic Geography,* Vol. 34, pp. 145-154.

Bhat, L.S. (1969): "Central Place Model as a Spatial Framework for Regional and National Planning in India", *Symposium* on the City as a centre of change in Asia, University of Hong Kong.

Brush, J.E. (1953): "The Hierarchy of Central Places in South Western Wisconsin", *Geographical Review,* Vol. XL, pp. 380-402.

Brush, J.E. (1968): "Spatial Pattern of Population in India Cities", *Geographical Review,* Vol. L (VIII), pp. 374-387.

Bracey, H.E. (1954): "Town as Rural Service Centre : An Index of Centrality with Special Reference to Somersit", *Transaction of the Institute of British Geographer,* Vol. 19, pp. 95-105.

Bromley, R.J. (1971): "Markets in Developing Countries—A Review", *Geography,* Vol. 6, pp. 124-132.

Cassettie, E. (1969): "Urban Population Density Pattern : An Alternative Explanation", *Canadian Geographer,* Vol. II, pp. 55-85.

Carol, H. (1960): "The Hierarchy of Central Function within the City", *Annals of the Association of Amercian Geographers,* Vol. 50, pp. 419-438.

Cohen, S.B. and George, K. Lewis (1967): "Form and Function in the Geography of Retailing", *Economic Geography,* Vol. 43, No. 1, pp. 1-42.

Clark, C. (1951): "Urban Population Densities", *Journal of the Royal Statistical Society,* Part IV, pp. 490-496.

Cohen S.B. and Applebaum, W. (1960): "Evaluating Store Sites and Determining Store Rents", *Economic Geography,* Vol. 36, pp. 1-35.

Davies, W.K. (1967): "Centrality and the Central Place Hierarchy", *Urban Studies,* Vol. 4, pp. 61-79.

Davies, R.L. (1972): Model of Retail Distribution", *Transactions of the Institute of British Geographers,* Vol. 57, pp. 59-82.

Decay, A. (1966): "A Probablity Model for Central Place Location", *Annals of the Association of American Geographer,* Vol. 56, pp. 550-568.

Dutt, A.K. (1969): "Daily Shopping in Howrah", *National Geographical Journal of India,* Vol. 12, pp. 163-167.

Dutt, A.K. (1967): "Shopping in Calcutta", *Town Planning Review,* Vol. 37, pp. 207-216.

Dutt, A.K. (1969): "Intra-city Hierarchy of Central Places, Calcutta, A Case Study", *Professional Geographers,* Vol. 21, pp. 18-23.

Gore, M.S. (1971): "Expanding Metropolis of Bombay", *Indian and Foreign Review,* Vol. 13, pp. 16-27.

Hoyt, H. (1961): "The Utility of the Economic Base Method in Calculating Urban Growth", *Land Economics,* Vol. 37, pp. 51-58.

Huff, D.C. (1963): "A Probabilistic Analysis of Consumer Spatial Behaviour", in William S. Decker (Ed.), "Emerging Concept in

Marketing", Proceeding of the Winter Conference of an American Marketing Association, Dec. (1962) Chicago, Cited by D.L. Thompson, "Future Direction in Retail Area Research", *Economic Geography,* Vol. 42, No. 1, p. 7.

Jefferson, Mark (1939): "The Law of the Primate City", *Geographical Review,* Vol. 29, pp. 226-232.

Johnston, R.J. (1966): "The Distribution of an Intra-Metropolitan Central Place Hierarchy in Melbourne", *Australian Geographical Studies,* Vol. 4, pp. 17-33.

Krishnan, V. (1932): "Fairs and Trade Centres of Madurai and Ramnad District", *Journal of Madras Geographical Association,* Vol. 7, No. 3, pp. 236-249.

Krishnan, V. (1940): "Trade Centre of Tinnevelly District", *Journal of Madras Geographic Association,* Vol. 15(3), pp. 272-276.

Kulkarni, K.M. (1975): "Shopping Pattern Near Greater Bombay—Suburban Railway Station - A Case Study", *Bombay Geographical Magazine,* Vol. XXIII, No. 1.

Lahiri, T.B. (1963): "Calcutta Metropolitan District", *Proceedings* of Indian Science Congress.

Lakdawala, D.T. and Sandesara, J.C. (1957): "Shops and Establishments in Greater Bombay", *Journal of the University of Bombay,* Vol. 24, No. 4, pp. 56-103.

Mattila, J.M. and Thompson, W.R. (1955): "The Measurement of the Economic Base of the Metropolitan Area", *Land Economics,* Vol. 31, p. 215.

Misra, H.M. (1977): "Economic Base of Indian City System", Paper read at the Regional Meeting of A.A.G. University of Michigan, U.S.A.

Morris, M.D. (1965): "The Emergence of an Industrial Labour Force in India", Berkeley University of California Press, Berkeley.

Mukerjee, A.B. (1957): "The Bi-weekly Market at Modinagar", *Indian Geographical Journal,* Vol. 32, pp. 35-42.

Nelson, H.J. (1953): "A Service Classification of American Cities", *Economic Geography,* Vol. 31, No. 3, pp. 180-230.

Newling, B.E. (1968): "The Spatial Variation of Urban Population Densities", Paper Contributed to the 21st I.G.G. Congress, New Delhi.

Nissel, H. (1977): "Greater Bombay : An Explanation of Structure, Function and Development", *Bombay Geographical Magazine,* Vol. 25, No. 1, pp. 1-23.

Palamaki, M. (1964): "The Functional Centre and Area of South Bothmia Finland", Fenna 88.1, Cited by Carter H. (1972): *The Study of Urban Geography,* Edward Arnold, London, pp. 81-82.

Pandey, M.C. (1975): "Masjid Bunder as a Wholesale Commercial Centre", Unpublished M.A. Dissertation, Department of Geography, Unviersity of Bombay.

Pathak, C.R. (1980): "Urbanisation Problem and Policies in India", *Indian Journal of Regional Science,* Vol. XII, pp. 71-90.

Pattanaik, N. (1953): "Weekly Market at Barpali", *Geographical Review of India,* Vol. 15, No. 1, pp. 1-20.

Proudfoot, M. (1937): "City Retail Structure", *Economic Geography,* Vol. 13, pp. 425-428.

Pred, A. (1976): "The Inter-Urban Transmission of Growth in Advance Economics, Empirical Findings Versus Regional Planning Assumption", *Regional Studies,* Vol. 10, pp. 151-171.

Ratcliff Richard, U. (1939): "Problem of Retail Site Location", *Michigan Business Studies,* Vol. IX, No. 1, Ann Arbor, University of Michigan, Cited by Cohen Saul B. and Lewis, George: "Forms and Function in the Geography of Retailing", *Economic Geography,* Vol. 43, No. 1.

Raymond, E. Múrhphy and Vance, J.C. (1954): "Delimiting the CBD", *Economic Geography,* Vol. 30, pp. 189-222.

Rees, P.H. (1971): "Factorial Ecology: An Extended Definition, Survey, Critique of the Field", *Economic Geography, Supplement* Vol. 47, pp. 220-233.

Reddy, S.N. (1961): "Vegetable Market and Regional Relationship of Hyderabad City", *Geographical Review of India,* Vol. 23, No. 3, pp. 24-40.

Reilly, William J. (1931): *Laws of Retail Gravitation",* New York, Pittsbury Publishers, New York.

Revali, K. (1977): "Pattern of Population Distribution in Bombay", Unpublished M.A. Dissertation, Department of Geography, University of Bombay.

Rogers, A. (1965): "A Stochastic Analysis of the Spatial Clustering of Retail Establishment", *Journal of the Amercian Statistical Association,* Vol. 60, pp. 1094-1103.

Rushton, G. (1911): "Postulates of Central Place Theory and the Properties of Central Place System", *Geographical Analysis,* Vol. 3, pp. 140-156.

Sahasrabudhe, S.S. (1975): "Occupational Structure of the Residential Areas of Greater Bombay—A Spatial Review", Unpublished M.A. Dissertation, Department of Geography, University of Bombay.

Sastry, G.S. (1981): "Retail Trade Structure of an Indian Metropolis, Bangalore", *Annals of the National Association of Geographers,*

India, Vol. 1, No. 2, pp. 57.73.

Sawant, S.B. (1978): "Shopping Area in the City of Poona", *Geographical Review of India,* Vol. 40, pp. 155-163.

Singh, O.P. (1975): "Central Place Region for Planning in Uttar Pradesh", *National Geographical Journal of India,* Vol. XXI, Part 2, pp. 98-107.

Singh, R.L. (1955): "Urban Hierarchy in the Umland of Benaras", *Journal of Scientific Research, Benaras Hindu Unviersity,* Vol. VI(2), pp. 181-192.

Sita, K. et al (1985): "Declining City Core of an Indian Metropolis—A Case Study of Bombay", U.G.C. Project Report, Department of Geography, University of Bombay.

Skinner, S.M. (1964): "Marketing and Social Structure in Rural China", *Journal of Asian Studies,* Vol. 24, pp. 195-228.

Smailes, A.E. (1944): "The Urban Hierarchy in England and Wales", *Geography,* Vol. 29, pp. 41-51.

Swaminathan M.S., (1979): "A Qualitative Appreciation in the Study of the Spatial Distribution and Functional Hierarchy of Shopping Centre in Coimbatore City", *National Geographicl Journal of India,* Vol. 25, pp. 231-254.

Tennants, R.J. (1961): "Population Density Pattern in Eight Asian Cities", Paper read at West Lakes Division Meeting of the Association of American Geographers.

Tewari, V.K. (1976): "Inter-State Disparities in the Level of Industrialization and per Capita Total Consumer Expenditure in India", Golden Jubilee Volume, the Indian Geographical Society, Madras.

Thompson, D.L. (1960): "Future Direction in Retail Area Research", *Economic Geography,* Vol. 4, pp. 1-18.

Wanamali, S. (1972): "Central Places and their Tributary Population : Some Observations, *Journal of Behavioural Sciences and Community Development,* Vol. VI (1), pp. 11-39.

Whebell, C.F.J. (1969): "Corridors : A Theory of Urban Systems" *Eskistis,* Vol. 28, No. 7, p. 168.

3. Government and Other Publications

Bombay Metropolitan Regional Development Authority (1985): "Basic Transport and Communication Statistics for Bombay Metropolitan Region".

Census of India (1961): *Maharashtra* Vol. X, Part I-C, Subsidiary Table.

Census of India (1971): *"Maharashtra,* Part IIA and B(i) General Economic Table, B1-B6.

Census of India (1981): *"Maharashtra"*, Part IIA and B(i) General Economic Table, B1-B6.

Census of India (1971): *Maharashtra;* General Population Table, Series II, Part IIA.

Census of India (1981): *Maharashtra,* General Population Table, Series II, Part IIA.

Census of India (1981): *Maharashtra,* Series 12, Part IIB.

Commerce Research Bureau (1985): "Economics of Bombay, *Financial Express,* May 31, 1985.

Central Road Research Institute (1983): "Planning of the Road System for Bombay Metropolitan Region," Vol. 1 and II, New Delhi.

Gazetteer of Bombay City and Island (1909), Vol. I, II and III.

Gazetteer of Bombay City and Island (1986), New Edition, Vol. I and II.

Kerkar A, (1981): "Report of the High Power Steering Group for Slums and Dilapidated Houses", Bombay.

Municipal Corporation of Greater Bombay (1980-81): "Administrative Report of the Municipal Commissioner for Greater Bombay."

Operation Research Group (1972): "Study on Economic Activity Profiles for Bombay, Baroda,

World Bank Report (1985): "The World Development Report", Oxford University Press.

INDEX